WOMEN IN AGRICULTURE

WOMEN'S HISTORY AND CULTURE
VOLUME 11
GARLAND REFERENCE LIBRARY OF SOCIAL SCIENCE
VOLUME 908

Women's History and Culture

Women in Agriculture
A Guide to Research
Marie Maman
and Thelma H. Tate

Women and Fundamentalism
Islam and Christianity
Shahin Gerami

Women and Technology
An Annotated Bibliography
Cynthia Gay Bindocci

The Higher Education of Women in England and America, 1865–1920
Elizabeth Seymour Eschbach

Lives of Women Public Schoolteachers
Scenes from American Educational History
Madelyn Holmes
and Beverly J. Weiss

Women in Agriculture
A Guide to Research

Marie Maman and Thelma H. Tate

Garland Publishing, Inc.
New York and London
1996

Copyright © 1996 by Marie Maman and Thelma H. Tate
All rights reserved

Library of Congress Cataloging-in-Publication Data

Maman, Marie.
 Women in agriculture : a guide to research / Marie Maman and
Thelma H. Tate.
  p.   cm. — (Women's history and culture ; v. 11) (Garland reference library of social science ; v. 908)
  Includes index.
  ISBN 0-8153-1354-3 (alk. paper)
  1. Women in agriculture—Bibliography.   I. Tate, Thelma H.   II. Title.
III. Series.   IV. Series: Garland reference library of social science ; v. 908.
Z7963.E7M316 1996
[HD6073.A29]
016.3314'83—dc20   95-41230
                                                                    CIP

Printed on acid-free, 250-year-life paper
Manufactured in the United States of America

We would like to dedicate this book to Ester Boserup for her pioneering work, *Women's Role in Economic Development.* She was the first to recognize how important women's role is in agricultural development, and she strongly encouraged further research in this area.

TABLE OF CONTENTS

I.	Introduction		ix
II.	Annotated Bibliography:		3
	1.	Historical Studies on Women in Agriculture	3
	2.	Women's Role in Agricultural Economic Development	33
		Africa	33
		The Americas	67
		Asia and Australia	82
		Europe	97
		General Studies	105
	3.	Sexual Division of Labor in Agriculture	119
	4.	Decision-Making on the Farm	151
	5.	Women's Role in Agricultural Policy Implementation	161
	6.	The Education of Women in Agriculture	179
	7.	Dissertations and Master's Theses	207
III.	Research Guide:		241
	1.	French Books and Articles	241
	2.	Bibliographies on Women in Agriculture	251
	3.	List of Journal Titles Frequently Publishing on Women in Agriculture	261
	4.	List of Journal Issues Related to Women in Agriculture	271
	5.	Electronic Resources and Indexes	273
IV.	Author Index		277
V.	Subject Index		287

INTRODUCTION

In what ways have women contributed to agriculture? To what extent have scholars addressed these contributions in the professional literature? What has been the impact of gender in agricultural policy and economic development? What is the status of gender equity in the division of farm labor and in agricultural education? Such questions are raised by students and researchers worldwide who seek documentation which focuses on these vital topics.

Our experiences as reference librarians at an academic library serving students and faculty at Cook College, the land-grant college of New Jersey and Douglass College, the largest women's college in the United States, have been that information on the many topics of women working in the field of agriculture is spread over many areas. Often it is necessary to use many sources to gather enough information to cover this topic with a degree of adequacy. The purpose of this bibliography is, therefore, to synthesize this unique widely dispersed information in one volume, to assist researchers, faculty, and students in expediting the research process. A faculty member preparing material for teaching a class or a student writing a paper in agriculture, anthropology, economics, history, or sociology will find this volume useful. It is an invaluable source for works related to women farmers, sexual division of labor in farming, economic, historical, or educational aspects of women on the farm.

We have included only information easily available at larger academic libraries or through inter-library loan. We have searched our own library catalog, and over 75 percent of the material we have included is available in our own library. We have also searched Research Library Information Network (RLIN), a database covering the book collections in the largest libraries in the United States, to add information we do not have at Rutgers, and we have used inter-library loan to secure this material. To locate information published in journals, we have searched many databases, such as Agricola and CAB, the Commonwealth Agriculture Bureau's database, *Historical Abstracts*, *America: History and Life*, *ERIC*, the Education Index online, *Sociological Abstracts*, the Wilson Indexes in *Social Sciences* and *Education Index*, *Abstracts in Anthropology*, *Women's Studies Abstracts* and *Studies on Women Abstracts* (see a description of these sources in the section under Research Guides).

The first part of the book includes the bibliography and is organized under broad topics, and the information in each chapter is listed alphabetically by author. The chapter on economics, where we found most of our information, is divided into major geographical areas and then alphabetically by authors. We have not included in this bibliography conference papers, symposiums, or reports from the many different organizations working with women in the fields of agriculture. A bibliography covering this literature was published in 1985 by the National Council for Research on Women and compiled by Shelly Kessler. The title is *Third World Women in Agriculture: An Annotated Bibliography*, and the author claims in the introduction that the purpose is to locate material on this subject available in the New York-Washington area. Ms. Kessler mentions in the introduction "how hard it was to locate information on this topic, and that traditional means of literature searching were not applicable to the subject." This is one of the reasons why we started searching the databases available in our library and compiled what we have found, but, contrary to that in Ms. Kessler's bibliography, the material we have listed is available in libraries throughout the United States.

The second part of the book is a research guide. Here, we have added a list of studies in French. Since a comprehensive bibliography with over 400 entries relating to South America was published in 1985 (Citation no 163), Spanish language entries have not been included in this work. The chapter on other bibliographies contains complete works on the subject, bibliographical articles in journals as well as works that devote chapters to the topic. Journals that publish on topics related to women farmers or women in agriculture, as well as whole journal issues on the same subject, are included. Finally, we added descriptions of electronic resources and indexes for sources not available in electronic format.

We thank our colleagues in the Mabel Smith Douglass Library for their support and encouragement, Phyllis Palfy for all of her help with inter-library loans, and Shirley Peck for getting books for us through the material delivery service at Rutgers.

Women in Agriculture

ANNOTATED BIBLIOGRAPHY

HISTORICAL STUDIES ON WOMEN IN AGRICULTURE

Women have become a subject worthy of serious research in the history of agriculture relatively recently, a change that parallels the general women's movements. Core bibliographies such as Catherine R. Loeb's and others[1] reflect the scarcity of unique sources on women in agriculture and highlight the need for further research in this area. While the role of women has been randomly intertwined in the history of agriculture from ancient times, there is now a determined need to document women's specific contributions as well as the conditions and impact of their work in this most vital area of life in global cultures.

In her book *Promise to the Land*, internationally recognized author Joan Jensen states: "Urban women held center stage and were the focus of the most exciting new research generated by the feminist movement of the 1970s."[2] A close review of the literature illuminates the correctness of this statement in that there are very few studies of this nature on rural women before the middle of the 1980s. Prior to this time, oral and local histories presented the most comprehensive records of rural women and their agricultural activities. Today, scholars recognize the important place rural women held in the history of their countries as presented through research, historical studies, and fiction on topics related to rural women's lives all over the world.

Clearly documented in all of these writings is the strong role that women have played as the backbone of their families and their communities, even though they were not given equal rights, pay equity, and deserved respect. The few men who have participated in writing the history of farm women have not highlighted the problems of unfair play that have existed for so long. While the role women have played in agriculture is still not adequately understood, appreciated, and documented, there is a growing body of literature which will be useful in raising consciousness and the visibility of women's role and will support demands for greater attention from politicians and scholars world-wide. This bibliography contributes to this development.

---

[1]Loeb, Catherine R. *Women's Studies: A Recommended Core Bibliography 1980-1985.* Colorado: Libraries Unlimited, 1987.

[2]Jensen, Joan M. *Promise to the Land, Essays on Rural Women.* Albuquerque: University of New Mexico Press, 1991, p.ix.

In preparing this part of the bibliography, many significant indexes and databases were searched, including *Historical Abstracts*,[3] *America: History and Life*.[4] Many of these sources are available via Dialog or CD-ROMs in larger public and academic libraries.

1. Adams, Jane. "Resistance to 'modernity': Southern Illinois Farm Women and the Cult of Domesticity." *American Ethnologist* 20:1 (1993): 89-113.

    Examines the twentieth-century transformation of southern Illinois farm women's lives and domestic habits by combining feminist historical analysis with Foucault's interpretation of "modernity" as the development of functional division of time and space in utilitarian, rationalized practices. Found that farm household transformations corresponded to transformations in the larger economy, while remaining distinct from developments in urban areas. Increasing engagement with commodity markets did not give rise to a strong domestic ideology or relegated women to the purely domestic sphere. Shows intensification of women's labor to supply subsistence for a larger labor force as men turned increasingly to commercial production. Includes statistics and a lengthy bibliography.

2. Allen, Ruth Alice. *The Labor of Women in the Production of Cotton*. New York: Arno Press, 1975.

    Deals with the life and work of women who lived on the cotton farms of Texas and who formed a considerable part of the group classified in the census of 1920 as "farm population," which contained 1,092,000 females. Provides many statistical tables on such topics as amount of field work done by black women, household work, periodical literature, books borrowed from libraries, average number of children under fifteen, and marital status. Includes questionnaires, references for further reading, and a detailed index.

---

[3]Historical Abstracts. Edited by Eric H. Boehm. Santa Barbara, California: American Bibliographical Center-Clio Press, 1955-.

[4]*America History and Life*. Edited by E.H. Boehm. Santa Barbara, California: American Bibliographical Center-Clio Press, 1964-.

3. Ankarloo, Bengt. "Agriculture and Women's Work: Directions of Change in the West 1700-1900." *Journal of Family History* 4:2 (1979): 111-120.

   Compares Scandinavian women's work in the agricultural fields to a study of three townships in Hamilton, Iowa, in the United States during the period 1870 to 1900, based on federal and state census data. Includes statistical tables about Sweden and the Hamilton studies as well as a short bibliography.

4. Atkeson, Mary Meek. *The Women on the Farm*. New York: Century, 1924.

   Reports results of interviews with over a thousand farm women at Field Days and Organization meetings in many parts of the country. Assesses nearly a thousand pieces of correspondence received in 1920 by *Farm and Home* and in 1922 by *The Farmer's Wife* from women in every state in the Union. Concludes that women on the farm are optimistic and do not contemplate quitting their work. Additional readings are cited in the appendix.

5. Baker, Gladys L. "Women in the U.S. Department of Agriculture." *Agricultural History* 50:1 (1976): 190-201.

   Reports activities of James Wilson, Secretary of Agriculture, in support of women near the end of nineteenth century. Describes the 1890s plight of Ernestine Stevens, librarian, in the transformation of her professional position in the Department of Agriculture to a clerical position with less pay (from $1800 to $1200). Highlights contributions of Dr. Louise Stanley, a graduate of Yale University, in the development and expansion of home economics as an important field for women in the Department, and where she was chief of Bureau between 1922 and 1943. Discusses salaries of over 4,000 women employed in the Department of Agriculture which increased from 19.9% to 34.9% between 1924 and 1943 but declined to 21.0% by 1947.

6. Bartley, Paula, and Cathy Loxton. *Plains Women: Women in the American West*. Cambridge, MA: Cambridge University Press, 1991.

Examines the sociological and physical challenges faced by women who settled in the Midwest, as presented in this Women In History series. Describes the contrast between whites and Native Americans on the Plains and particularly the vital roles of Native American women in that environment.

7. Beers, Howard W. "Portrait of the Farm Family in Central New York State." *American Sociological Review* 2:5 (October 1937): 591-600.

Compares vital, geographically isolated, economically self-sufficient, dominant father and obedient wife in pioneer American family with twentieth century families. Reveals the family of the 1930s as being smaller, less family oriented, with less obedient children as well as larger economical and sociological influences impacting agriculture and family traditions.

8. Binni-Clark, Georgina. *Wheat and Women.* Toronto: University of Toronto Press, 1979.

Tells the story of an English woman writer settling as a wheat farmer on the Canadian prairie in Saskatchewan. Autobiographical description of her experiences with her brother as she became a successful independent farmer. Describes her life from poverty to prosperity, and the differences between being a farmer's wife and a woman farmer. Offers advice to other women who would like to make a living in agriculture.

9. Borish, Linda J. *The Last Of the Farm: Health, Domestic Roles, and the Culture of Farm Women in Hartford County, Connecticut 1820-1870.* Ph.D. Dissertation, University of Maryland, 1990.

Examines the health of farm women in Hartford County, Connecticut, 1820-1870, and issues of health and physical well-being as part of a large male-female conflict over rural life. Explains levels of power exercised by women and men with respect to their interpretation of the nature of farm living along gender lines. Outlines the exposure of young rural women to city culture and new options. Discusses the exodus of young women from rural to urban areas, which shocked female agriculturists and reformist males who asserted that farm females' terrible quality of life needed to be redressed.

Discusses the impact of reformist efforts to promote physical recreation and sport, improved mental culture, home embellishments, domestic labor, health-saving devices, as well as participation at agricultural fairs as antidotes to the perceived stagnation and ill health of the rural women. Includes an extensive list of references.

10. Brady, Marilyn Dell. "Populism and Feminism in a Newspaper by and for Women of the Kansas Farmers' Alliance, 1891-1894." *Kansas History* 7:4 (1984-1985): 280-290.

Describes domestic and political concerns of farmers as well as famous and obscure writings of women in the radical group of the Kansas Alliance Movement published in the *Farmer's Wife*, a monthly newspaper published in Topeka, Kansas, from 1891 to 1894. Highlights contributions of Emma D. Pack, editor and co-owner of the paper; Bina Otis, wife of Populist Congressman John Otis; and Fannie McCormick, the female foreman of the Kansas Knights of Labor.

11. Brown, Minnie Miller. "Black Women in American Agriculture." *Agricultural History 50:1 (1976): 202-212.*

Identifies major contributions and experiences of black women in the development of American agriculture through an in-depth review of levels of education. Looks at the social effects of the plantation system on the unskilled black laborers. Explains major role of southern black women as wage laborers and their limited role as owners (only 8% were farm owners in 1940). Discusses the positive and negative impact of the Agricultural Adjustment Act of 1933 which raised and stabilized farm income, reduced acreage in the main crops requiring labor, drastically increased the number of Black sharecroppers and tenants who were "pushed off" the land. Includes a short list of references.

12. Buffalohead, Priscilla K. "Farmers, Warriors, Traders: A Fresh Look at Ojibway Women." *Minnesota History* 48:6 (1983): 236-244.

To understand the dimensions of women's status in the historical culture of the Ojibway Indian people of Minnesota and the upper Great Lakes region, a critical evaluation of the information provided in primary and secondary historical sources is necessary. These sources provide biased and often contradictory images of native women as well as valuable insight into their life. Two pictures of Ojibway women can be found: one portraying them as slaves of the men, the other depicting them in a far more dynamic role in the political, economic, and social life of their communities. It is necessary to understand major trends in Western thought about women as well as the history and culture of the Ojibway people in order to sort out the historical writings and arrive at a semblance of truth. A list of references is included.

13. Carney, Judith, and Michael Watts. "Disciplining Women? Rice Mechanization, and the Evolution of Mandinka Gender Relations in Senegambia." *Signs: Journal of Women in Culture and Society* 16:4 (Summer 1991): 651-681.

Details a 150-year process of agrarian change in a peasant society that experienced repeated efforts to intensify the labor process. Describes the female farming system in the nineteenth century as part of a wider process to intensify food production in general and rice production in particular. Traces in detail the series of strategies used to reduce reliance on imported rice as well as efforts implemented after World War II to mechanize double cropping under irrigated conditions. Includes over 80 references in the footnotes.

14. Cohen, Marjorie Griffin. *Women's Work, Markets, and Economic Development in Nineteenth-Century Ontario.* Toronto: University of Toronto Press, 1988.

Demonstrates how the neglect of analysis of household economy created a narrow view of the economic development in Ontario in the nineteenth century. Offers an alternative to the usual ideas about the impact of industrialization on women. Argues that women's involvement in the industrialization process took a different form than in older societies which were industrialized first. Shows that the productive relations of the household are critical to analyses of the nature of economic development. Studies the staple-exporting economy in Ontario in the nineteenth century, women's participation, and the changing conditions of women's work in dairying. Describes women's paid work and the transition to industrial capitalism during the years 1850 to 1911. Develops a new dimension to the study of women's labor history by exploring the roots of economic development. Includes statistical tables, extensive bibliography, and an index.

15. Craig, Lee A. "The Value of Household Labor in Antebellum Northern Agriculture." *The Journal of Economic History* 51:1 (March 1991): 67-81.

   Reviews contributions of labor by age, sex, and region in antebellum northern agriculture. Discusses household labor by children between the ages of seven and twelve and contributions of teenage females in the old Northwest. Teenage boys, adult men, and adult women made their largest contribution in the Northeast, contrary to the widely held view that children contributed more in the least settled regions.

16. Ericksen, Julia, and Gary Klein. "Women's Role and Family Production among the Old Order Amish." *Rural Sociology* 46:2 (1981): 282-296.

   Examines ways in which the productive roles of Amish women help maintain Old Order Amish society and the way these roles vary with women's position in the life cycle. Confirms that in a labor intensive agricultural society, women perform important economic roles both in terms of their subsistence activity and in their reproduction of children to provide the hands that maintain the system.

17. Evans, Barbara. "Yours for a Square Deal: Women's Role in the Saskatchewan Farm Movement and Early CCF." *Canadian Dimension* 21:3 (1987): 7-12.

   Explores the farm women's organization from 1913 to 1933 of the Saskatchewan farm movement when it went from the politics of self-reliance to the politics of socialism. Describes contributions of Francis Marian Beynon, the women's editor of the *Grain Grower's Guide,* who organized the first convention of the Saskatchewan Women Grain Grower's Association in February 1913. The association provided a forum for women to exchange ideas and sow the seeds of a provincial network of rural women. In 1916 prairie women became the first women in Canada to obtain the vote, the result of a concerted campaign which united both rural and urban women. The underlying philosophy of the association was cooperation, but even as early as 1915, individual farm women had discussed the merit of socialism. The farm movement became more political during the 1920s and 1930s, and the farm women were encouraged to use their leisure time in productive study. "Learn, learn" became the slogan of the day. In 1926 a new organization, the United Farmers of Canada (Saskatchewan Section) UFC(SS) was formed. The separate board of the Women Grain Growers was abolished, and women were represented with the election of a women president, who became a member of the central executive, as well as two women directors. By 1930, all officers were elected by both men and women delegates at the same convention, and women were nominated as president and vice-president of the main organization but not elected.

18. Fairbanks, Carol, and Sara Brooks Sundberg. *Farm Women on the Prairie Frontier: A Sourcebook for Canada and the United States.* Metuchen, NJ: Scarecrow Press, 1983.

Introduces students, teachers and general readers to historical and literary materials of the grasslands of Canada and the United States. Presents four essays introducing readers to the land and the people, the history and the fiction. The second part is an annotated bibliography of history, fiction, nonfiction, and women's fiction.

19. Fairbanks, Carol. *Prairie Women: Images in American and Canadian Fiction.* New Haven CT: Yale University Press, 1986.

Presents an historical account of prairie women in the United States and Canada, "seeing with new eyes, and entering old text from a feminist critical perspective." Analyzes the way women writers have described the experiences of prairie women and how they described the "new" land, the land that was new to the pioneers but old and familiar to the native people. Includes illustrations, a bibliography, and an index.

20. Fairbanks, Carol, and Bergine Haakenson. *Writings of Farm Women 1840-1940: An Anthology.* New York: Garland Publishing, 1990.

Presents stories by farm women relating their experiences with various farming techniques, flower gardens, animals, religious life, politics, and destruction of crops by grasshoppers and drought. Describes their experiences as they occurred, paying limited attention to more painful situations while highlighting "dignity of women who worked hard and developed self-confidence," as stated by the editors in the preface. Includes a short index.

21. Faragher, John Mack. "History from the Inside-Out: Writing the History of Women in Rural America." *American Quarterly* 33 (Winter 1981): 537-557.

American rural farm women, representing a majority of the female population well into the twentieth century, are among the most underrepresented of all Americans in the standard histories. Some rural women such as Native Americans, Afro-Americans, Chicanas, Mexicanas, and Hispanics, have suffered a historical silence because of the social conditions that kept them totally illiterate. Even white female illiteracy was commonplace until the second half of the nineteenth century. In spite of all this there is a surprisingly large amount of information. First, the basic documents of social history: the census data, probate materials, which traces individual families in the official records. Includes reminiscences, autobiographies, and oral testimonies transcribed by family members and antiquarians long before historians seriously considered oral history as important, and then there are some letters and diaries. "The problem has been blindness, not inarticulateness," says the author. Women's history should not be separated from general history, economic changes, or changes in politics and ideology that are central to understanding women's historical experience.

22. Farnsworth, Beatrice, and Lynne Viola, eds. *Russian Peasant Women*. Oxford: Oxford University Press, 1992.

Looks at Russian history in new and different ways and questions assumptions that the Revolution was a unique milestone in women's history. Treats peasant women before the Revolution describing their lives and family roles. Covers events after the Revolution and their place in Soviet agriculture. Shows the extent to which Soviet agriculture became dependent on female labor following migration of the young and the men to the urban sectors throughout the Soviet Union. Women remained in the countryside and took over the hard labor in agriculture; very few held managerial positions or were employed in skilled and mechanized work in the collective-farm system. Demonstrates that state-initiated modernization did not alter the position of women in agriculture. Includes statistical tables and references.

23. Fink, Deborah. *Agrarian Women: Wives and Mothers in Rural Nebraska 1880-1940.* Chapel Hill: The University of North Carolina Press, 1992.

An anthropologist, the author returns to Boone County in her home state of Nebraska to gather data on women in rural Nebraska in the late nineteenth and early twentieth centuries. Discusses the difficulties of getting women to talk about the past, because it contained experiences that were heartbreaking and humiliating. Covers land ethics, women's work, fertility, rural violence, agrarianism, and current farm politics. Offers a history of agrarianism and rural development in Boone County and discusses the sample of the women who were interviewed. Includes a bibliography and an index.

24. Fink, Deborah. *Open Country, Iowa: Rural Women, Tradition and Change.* Albany: State University of New York Press, 1986.

Deals with an anthropological study of a rural community in Iowa in the years following World War II. Derives data from in-depth interviews with 43 women and 5 men, census statistics, and historical records from churches and the county extension office. Discusses the changes that took place in women's control of egg production, which provided supplementary income to farm families during the years 1925 to about 1974. Discusses the impact of technological changes in poultry and egg production, which forced women to take low-paying jobs in the nearby towns and changed the nature of farming in rural Iowa. Includes a bibliography and an index.

25. Flora, Cornelia Butler, and John Stitz. "Female Subsistence Production and Commercial Farm Survival among Settlement Kansas Wheat Farmers." *Human Organization* 47:1 (Spring 1988): 64-69.

Discovers the participation of women in the emerging farm system in Ellis County, Kansas, from the period 1885 to 1905. Focuses on data gathered from the annual Kansas Agricultural Census and shows how precollected data on household and farm enterprises, combined with extensive historical analysis of diaries and contemporary accounts, can be used to piece together women's contribution to capital accumulation. Used a computer based parametric statistical technique.

26. Flora, Cornelia Butler and Jan L. Flora. "Structure of Agriculture and Women's Culture in the Great Plains." *Great Plains Quarterly* 8:4 (1988): 195-205.

Analyzes farming practices by two different ethnic groups on the plains. The U.S.-born farmers developed a strategy that can be classified as entrepreneurial. They focused on cash crops, had minimal production of subsistence crops, and because they depended more on hired labor, were more prone to substitute capital for labor when possible. Women's activities were focused in the reproductive areas, involving both housework and cultural activities, and education. In contrast, the more conservative German-born farmer employed a yeoman farm strategy, only expanding when necessary to set up a son in farming and only cautiously investing in machinery. Labor was seldom hired since family labor was readily available. The work of women and young girls in subsistence production was a key but little-valued part of this emerging farming system.

27. Fredricks, Anne. "The Creation of 'Women's Work' in Agriculture: The Women's Land Army during World War II." *The Insurgent Sociologist* 12:3 (1984): 33-40.

The history of the Women's Land Army provides an example of the way in which women's labor power was manipulated and utilized under particular historical circumstances. Women were not normally hired for these jobs because of their sex, and after the war women were expected to return to their narrow pre-war employment options or move out of the labor force altogether to make room for men. Over two million women participated in the Women's Land Army, and it was extremely successful. This concrete illustration of sex/gender and labor force relations points directly to the more complex issues surrounding the sexual division of labor in society. A list of references is included.

28. Gebby, Margaret Dow. *Farm Wife: A Self Portrait,* 1886-1896. Edited by Virginia E. McCormick. Ames, Iowa: Iowa State University Press, 1990.

Discusses Margaret Dow Gebby as an ordinary woman whose name was unknown beyond her family and community. Her husband owned a grain and livestock farm near Bellefontaine, Ohio, and her three sons and mother lived with them during the period covered by her diaries. She was unusually faithful in keeping daily records of family activities of farm and household income and expenses for more than a decade. Gives a noteworthy perspective of a woman's viewpoint of farm work and daily life, a family record that reveals dynamics of changing relationships as youth become adults and grandparents die, and details financial records over a decade of fluctuation that reflects family adjustment to agricultural prosperity and depression and a record of the technological innovations adopted by one farmer. Discusses leisure time, holidays, religious activities in the community as well as transportation and communication in this part of the country. Includes a short glossary and an index.

29. Gloss, Molly. *The Jump-Off Creek.* Boston: Houghton Mifflin Company, 1989.

Focuses on hardship, loneliness, freedom, decision-making, and self-rule which the heroines in the tale had been denied as daughters or wives. Based on published and unpublished diaries, letters and journals of women who settled in the West.

30. Hagood, Margaret Jarman. *Mothers of the South: Portraiture of the White Tenant Farm Woman.* New York: Arno Press & The New York Times, 1972.

Working in a triple role as mothers, housekeepers, and field laborers with no training, the mothers of the South produce and care for families and crop. The achievements here portrayed can only be interpreted as evidence of the existence of inherent quality, vitality, and endurance in the people, as well as certain facilitating factors and enhancing values of rural life. Focuses on the mothers and how their lives are affected by the occupation of their husbands and the farms which the tenant husbands are able to rent. Presents case studies from the Piedmont section of North Carolina, an area of about

8,000 square miles with a population of 700,000. Analyzes the findings in the final two chapters. Includes an index.

31. Haney, Wawa G., and Jane B. Knowles, eds. *Women and Farming: Changing Roles, Changing Structure.* Boulder, Colorado: Westview Press, 1988.

    Includes twenty essays from the second National Conference on American Farm Women in Historical Perspective. Relates changes in the structure of U.S. agriculture to the role of women in agriculture. Discusses the impact of social and economic changes on farm women; farm women's economic roles; farm women and resource control; farm women in comparative and historical perspectives; and farm women's community and political roles. Includes a bibliography, index, and biographical data on the authors.

32. Hargreaves, Mary W. M. "Homesteading and Homemaking on the Plains: A Review." *Agricultural History* 47 (1973): 156-163.

    Reviews three books: Faye Cashatt Lewis, *Nothing to Make a Shadow.* Ames: The Iowa State University Press, 1971; Sarah Ellen Roberts, *Alberta Homestead: Chronicle of a Pioneer Family.* Austin: University of Texas Press, 1971; Walker D. Wyman, *Frontier Woman: The Life of a Woman Homesteader of The Dakota Frontier, Retold from the Original Notes and Letters of Grace Fairchild.* University of Wisconsin: River Falls Press, 1972.

33. Hargreaves, Mary M.W. "Women in the Agricultural Settlement of the Northern Plains." *Agricultural History* 50 (1976): 176-189.

    Reviews the hardship of women during the settlement of the North American Plains. Includes references.

34. Harris, Evelyn. *The Barter Lady: A Women Farmer Sees It Through.* Garden City, N.Y.: Doubleday, Doran Inc., 1934.

Gives a detailed account of a widow of nine years and how she manages to keep the family of five fatherless children, plus three farms with a big mortgage - on nothing. Goes through the year, month by month, and relates the Barter Lady's appearances before three Judges of the Orphans Court, "which function only when the female family members are left alone. Reveals the lack of concern about the manner in which male members of the family manage money, the farm, or the real estate. Focuses on legal obstacles encountered by females in Maryland and South Carolina. Discusses differences in social treatment of females based on gender, in this tale that reads like a novel. States the truth in a very matter-of-fact way, without complaining.

35. Hill, Bridget. "Female Servants in Husbandry" in Bridget Hill, *Women, Work, and Sexual Politics in Eighteenth-Century England.* New York: Basil Blackwell Inc., 1989, pp. 69-84.

Asserts the difficulties of determining with a degree of accuracy the number of female servants in husbandry in eighteenth-century England. Describes their situation which sometimes made them particularly vulnerable to exploitation and abuse. Highlights social development among women of the lower classes with respect to entry into farm service, marriage to cottage or small farmers, and farm ownership. Discusses decline of service in husbandry in the middle of the century when women lost not only the training and a productive role in agriculture but also lost experiences in leaving home for a training shared with the opposite sex, not to be regained for many years.

36. Holmes, Francis W. and Hans M. Heybroek. *Dutch Elm Disease, The Early Papers: Selected Works of Seven Dutch Women Phytopathologists.* St. Paul, Minnesota: APA Press, 1990.

Explores the foundation of knowledge of the Dutch Elm disease laid by seven women scientists in the Netherlands. Discusses the discovery and location of the disease in 1919, which had already covered most of Belgium, the Netherlands, and part of France. Describes the research by these women and gives a short biography and photos on each scientist: Barendina Gerarda Spierenburg (1880-1967); Marie Beatrice Schwarz (1898-1969); Johanna Westerdijk (1883-1961); Christine Johanna Buisman (1904-1988); Johanna

Catharina Went (1905-); Louise Catharina Petronella Kerling (1900-1985); Maria Sara Johanna Ledeboer (1904-1988).

37. Janiewski, Dolores. "Women and the Making of a Rural Proletariat in the Bright Tobacco Belt, 1880-1930." *Insurgent Sociologist* 10:1 (1980):16-26.

    Focuses on tobacco farms of the North Carolina Piedmont called "a factory without walls." Explores the rural roots of a process that led women into the tobacco and textile factories of Durham and other industrial communities. Examines why women in greater numbers than men and black women in greater numbers than white women joined the labor forces flowing from the fields to the factories. Since it is much harder to trace women than men through standard historical demographic methods, the author relies on a mixture of oral history testimony and an analysis of samples taken from the 1880 and 1900 manuscript census for Durham and neighboring counties to illuminate the sex, class, and race relationships in which rural women participated. Discusses women who entered the growing industrial centers of the Piedmont who had already become proletarianized before they began their journey into Durham. Relates how members of black and white tenant households had become permanently entrapped in a class of landless farmers, where women's labor did not receive the same rewards in money or respect as men's. Women found that work in a factory community was the only way by which they could support themselves and contribute to their families.

38. Jeffrey, Julie Roy. *Frontier Women: The Trans-Mississippi West 1840-1880*. New York: Hill and Wang, 1979.

    Focuses on the thousands of white American women going to the trans-Mississippi West in the decades of heavy migration. Addresses questions about special economic or political opportunities, the impact on family norms, the role of women in the process of mediation between culture and environment in building new communities on the frontier and draws evidence from over two hundred women's journals, letter collections, and interviews. Includes an extensive bibliography and an index.

39. Jeffrey, Julie Roy. "Women in the Southern Farm Alliance: A reconsideration of the Role and Status of Women in the Late Nineteenth-Century South." *Feminist Studies* 3:1-2 (1975): 72-91.

   Discuses the slogan of the Southern Farmers' Alliance: "equal rights to all, special privilege to none" and the manner in which woman was admitted into the organization as the equal of her brother. Describes a case study of the North Carolina Farmer's Alliance; shows that the Alliance also offered numerous rural women the rare privilege of discussing economic and political issues with men and of functioning as their organizational equals. Highlights official positions of women as secretary with responsibilities of recording minutes of meetings and receiving all monies due and making speeches that were published in the *Progressive Farmer*, the Farm Alliance official paper.

40. Jensen, Joan M. "Butter Making and Economic Development in Mid-Atlantic America from 1750 to 1850." *Signs: Journal of Women in Culture and Society* 13:4 (1988): 813-829.

   Looks at a specific agricultural task in butter making to document and evaluate the labor of women whose work is absent from census and journals. This study covers butter making in the Philadelphia area between 1750 and 1850, and it reveals that butter making was dominated by women in the late eighteenth century. Butter became more profitable than agricultural commodities produced by men, such as grain and livestock. With an estimated population of more than 42,000 in 1790, Philadelphia provided an important butter market and stable demand and price. Women's butter making became a valuable asset to the economy on the farm. Discusses the history, changes in tools, and techniques for butter-making.

41. Jensen, Joan M., and Gloria Ricci Lothrop. *California Women: A History*. San Francisco: Boyd & Fraser Publishing Company, 1987.

   Bring to the reader the variety of the rich cultural history created by Californian women. Describes Native Californian women, the Hispanic Californians, the westering Euro-American women, Asian and African women. Discusses different agricultural experiences of each group in chapter 1. Focuses on the urban frontier, the progressive Era, 1900-1924, the decade of discontent and between the Wars, and the contemporary women, 1960-

1980. Points out efforts of Californian women's search for equality and justice, which has been guided by a vision of a multicultural, gender-equal society.

42. Jensen Joan M. "Cloth, Butter and Boarders: Women's Household Production for the Market." *Review of Radical Political Economy* 12:2 (1980): 14-24.

Presents an historical framework for analyzing American women's household production for the market during the late eighteenth century and the beginning of the nineteenth century. Argues that household productions, such as making of cloth and butter and taking in boarders, counted as a crucial economic factor in both urban and rural families. When this mode of production ended, women moved into wage labor, which shifted the focus of their work from the home to the workplace.

43. Jensen, Joan M. "I've Worked, I'm Not Afraid of Work: Farm Women in New Mexico 1920-1940." *New Mexico Historical Review* 61:1 (1986): 26-52.

Rural women's history is accessible through census data, agricultural extension records, and especially oral history and is a particularly rich field of study. Describes the work of New Mexico farm women in family and community. It is confined primarily to the Hispanic and Anglo American majority, and Native American women and black women are not discussed in this study. The areas covered include Bernalillo, Dona Ana, Union Rio Arriba, Sante Fe, Taos, and Valencia counties. The majority of each group was desperately poor, and all New Mexico farm women were affected by certain economic facts, such as the changing world and domestic markets after World War I. The depression that spread out from urban areas and combined with one of the worst droughts in the history of the Southwest caused much devastation among the poor. During these decades, women's work on the farm was essential and crucial for the survival of the family.

44. Jensen, Joan M. *Loosing the Bonds: Mid-Atlantic Farm Women, 1750-1850*. New Haven: Yale University Press, 1986.

Documents farm women's work and lives in the Philadelphia area during the years 1750 to 1850. Employs wills, inventories, poorhouse and church records, to explain women's contributions to farming, to household work, and to the community. Focuses on the commercial sphere where women directly shared in economic development. Portrays women as shapers of the emerging commercialization of the farms. Discusses butter making as a cottage industry that provided an important source of income to farm families during a period of economic transition. Includes studies of women's activities in religion, education, and reform. Includes an appendix with tables on household inventories from Chester and Castle Counties in Pennsylvania and statistics on butter prices in Philadelphia from the years 1748 to 1849 as well as notes and an index.

45. Jensen, Joan M. *Promise to the Land: Essays on Rural Women.* Albuquerque, NM: University of New Mexico Press, 1991.

Describes historical conditions wherein women have seldom had control of the land on which they have labored: in the nineteenth century, when women married they lost their civil rights, including the right to own land in their own name; and the twentieth century, long after legal disabilities were removed in most states, social custom continued to restrict ownership and control by women. Highlights federal tax laws prior to 1982 which discriminated against women, i.e., if the husband died first the widow had to pay tax on the entire value of the estate belonging to him, but if the wife died first the husband had to pay no federal tax.

Includes three autobiographies of women from the past, oral history and iconography. Shares her own family history, because she feels we have to begin study the history of the farm women with our own past. Includes a chapter on the Native American women and the New Mexico farm women and summarizes the role of farm women in American history. Comments on historians' perspective that rural women had a more political active role than traditional history has reported. Accepts responsibility for her role as a historian and provides historical information to assist in formulating new policies for the farm women today.

46. Jensen, Joan M., ed. *With These Hands: Women Working on the Land.* Old Westbury, NY: The Feminist Press, 1981.

Explains how difficult it was to find documents on women working on the land, as most rural women left no records of their lives because they were not literate. Discusses lack of written language among Native American women and prohibition of literacy and written language among African-Americans. Asserts developments of the nineteenth century that provided opportunities for many Hispanic women to become literate in Spanish and English. Relates a similar development for European immigrant women who left very few literary traces before the 1980s. Examines all periods of American agricultural history. Includes references and an index.

47. Jones, Lu Ann and Nancy Grey Osterrud. "Breaking New Ground: Oral History and Agricultural History." *The Journal of American History* 76 (September 1989): 551-564.

Reflects on oral history is part of the continuous recreation of individual, familial, and collective identity and on the methodological issues raised by the involvement in both personal and political process as historians become more aware of different ways of reconstructing the past. Illuminates profound structural changes in American agriculture during the past half century through interviews that neither memorialize nor indict the present but rather allow farm people to describe and interpret their experience of change. Illustrates the irony that success in farming may lead to eventual failure. A combination of several factors led a North Carolina family and their children from the land and pushed them out of farming. Factors which contributed to this outcome included educational opportunities, comfortable standard of living, freedom of choice, and the high risk of heavy responsibilities that commercial farming entailed.

48. Kinnear, Mary. "Do You Want Your Daughter To Marry a Farmer? Women's Work on the Farm, 1922." *Canadian Papers in Rural History* 6 (1988): 137-153.

Uses two main sources to examine farm women's work in Manitoba. 1) a survey of the work of members of the United Farm Women in Manitoba, a four-part survey dealing with socioeconomic standing, size of the household, distances from social and community services, and availability of transportation, telephone, etc. as well as indices of labor inside the farm house and its immediate vicinity; 2) testimony of the farm women who responded to an essay competition in the *Grain Growers Guide* on the topic, "Should My Daughter Marry a Farmer?"

49. Kurian, Rachel. *Women Workers in the Sri Lankan Plantation Sector: A Historical and Contemporary Analysis.* Geneva, Switzerland: International Labor Office, 1982.

    Examines the nature and evolution of the plantation system in Sri Lanka. Analyzes the various activities undertaken by the female workers and the economic implications for the plantation sector and the country. Includes an historical background of the Sri Lankan plantation and a discussion of the sociological structure of plantation labor and its implications for women. Examines the production and reproduction activities of women, income and expenditure, and welfare facilities. Includes a bibliography.

50. Maret, Elizabeth. *Women on the Range: Women's Roles in the Texas Beef Cattle Industry.* College Station: Texas A&M University Press, 1993.

    Asks "Are women represented in the Texas beef cattle industry and in what way are they represented?" Data are drawn from the analysis of government census statistics, survey research, and participant observation. Give historical information from 1800 to 1950 as well as information about the changing industry and roles of the ranching women in the future. Includes a survey questionnaire, bibliography, and an index.

51. Martelet, Penny. "Women's Land Army, World War I" in Mabel E. Deutrich and Virginia C. Purdy, eds. *Clio Was a Woman: Studies in the History of American Women.* Washington, D.C.: Howard University Press, 1980, pp.136-146.

    Explores the Land Army, part of a women's movement, which helped to open a door for the partial acceptance of hard physical labor for women and crashed class barriers for women from various walks of life as they lived together in cooperative communities. Discusses the idea of mobilizing women as agricultural workers in Great Britain, which resulted in a 40,000 British Women's Land Army by the end of World War I. Highlights two professors of geology at Barnard College, Ida H. Olgivie and Delia W. Marble, who jointly owned a farm in upstate New York. They first promoted the idea in the United States and turned their farm over to a dozen interested women students at Barnard. Describes the successful experiment that resulted in Professor Olgivie spending a year promoting the idea across the United States. Discusses the impact of the Land Army which provided women with war work, and opened new avenues of employment for women.

52. Marti, Donald B. *Women of the Grange: Mutuality and Sisterhood in Rural America, 1866-1920.* Westport, Connecticut: Greenwood Press, 1991.

Outlines historical development of the Grange, a secret fraternal organization that first appeared in 1866, flourished during the 1873 depression, declined abruptly after 1875, and slowly revived in the 1880s. Compares the first and second Grange movements: earlier members were midwestern, but the second Grange movement that began in the 1880s was predominantly eastern. Describes the organization's mission which fostered cooperative enterprises, called for economic reforms, and tried to improve farmers' social and intellectual lives by bringing neighbors together for enjoyable and stimulating meetings. It was meant to serve farmers in particular, but it was never been an organization of farmers exclusively. Observes women's participation in Grange activities, their role on committees, and the extent to which men and women could participate together in certain ceremonies. Includes extensive list of references.

53. McCurry, Dan C., ed. *Women in the Fruit-Growing and Canning Industries in the State of Washington: A Study of Hours, Wages and Conditions.* Washington, D.C.: U.S. Department of Labor, Women's Bureau, 1926.

Focuses on a study conducted by the Women's Bureau of the Department of Labor to investigate the conditions under which women and children were employed in outdoor industries, especially on fruit ranches in the State of Washington. Investigates two kinds of work: outdoor employment of women picking berries and fruit; and indoor employment of women canning vegetables, fruits, salmon and clams. Includes information about 219 canning, preserving and food packing ranches visited as well as 63 related establishments. Provides detailed information about housing, working conditions, time worked, and earnings of workers. Also includes information on occupational history, labor turnover, industrial accidents, and diseases of workers. Includes many statistical tables.

54. McMurray, Sally. "Women's Work in Agriculture: Divergent Trends in England and America, 1800 to 1930." *Comparative Studies in Society and History* 34:2 (1992): 248-270.

Looks into the history of women's work in English cheese dairying and how the sexual division of labor in American farm-based cheese dairying diverged in significant ways. Contrasts the value of such material as dairying manuals, travel accounts, Board of Agriculture reports, agriculture journals, and archival records in building a multifaceted account of the history of dairying with census data which recorded that not one single person practiced the occupation of dairy maid, even in the years of the mid nineteenth century, when thousands of women were making tons of butter and cheese. Reports on the success of a cheese dairy farm in England which depended upon a social arrangement in certain tasks that were consistently allocated to different sexes: men worked in the fields, tended, herded, and sheltered the cows, while the women made the cheese and reared the calves. Highlights the division of labor for outdoor and indoor tasks, except for milking which was shared. Focuses on many English dairy farms where women exercised a substantial amount of control of their work and, because of their cheesemaking skills, women often enjoyed an elevated status in the family and the community.

Discusses the history of British colonists who brought cheesemaking methods to America, but it was not until the period after 1825 that cheese dairying emerged as a specialty in American agriculture. Men were head cheesemakers to the greatest extent in the districts of central New York, New England, and Ohio, where cheese dairy farms produced thousands of pounds of cheese annually. Americans discovered that cheese offered better returns than the old staples of wool and grain; men studied cheesemaking methods with an eye to efficiency and shipping quality while women sought less arduous and more prestigious work such as teachers in the rapidly growing public school system. Points to the contrast between the experience of women in English and American cheesemaking to suggest that the pace of defeminization depends upon more than the farming specialty.

55. Mitra, Manoshi. "Women in Colonial Agriculture: Bihar in the Late 18th and the 19th Century." *Development and Change* 12 (1981): 29-53.

Evaluates the historiographical trends particularly in Indian history to illuminate the history of women in agriculture and traditional handicrafts in Bihar, a state of Eastern India, under colonial rule. Provides an overview of traditional historiography and women's history and briefly recounts the economic history of the area in the given period; presents some well documented work patterns of women who, along with their families, occupied a low socio-economic status in an agrarian society and shows how their status was adversely affected by British domination. Includes a short bibliography.

56. Moore, Henrietta, and Megan Vaughan. "Cutting Down Trees: Women, Nutrition and Agricultural Change in the Northern Province of Zambia, 1920-1986." *African Affairs (G.B.)* 86:345 (1987): 523-540.

Focus on women's labor which highlights previously ignored aspects of agricultural production and identifies problems of nutrition and food supply within a long-term historical perspective as a way of deconstructing some of the myths about slash and burn (citemene) agricultural system in northern Zambia. The neglect of village gardens in the discussion of agriculture in the Northern Province distorted the picture. The problem of female headed households requires a more complex analysis in which the absence of male labor is merely one factor.

57. Osterud, Nancy Grey. *Bonds of Community: The Lives of Farm Women in Nineteenth-Century New York.* Ithaca, N.Y.: Cornell University Press. 1991.

Documents the structure through which rural women were defined through their relationship with men. The kinship system identified them as wives or mothers, daughters, or sisters, and they gained access to land, the most important resource in an agricultural society, only through husbands and sons, fathers, or brothers. However, their labor remained integral to farm family economics as agriculture commercialized, and farm women contributed to commodity production in concrete and visible ways. In some parts of Pennsylvania and New York, the women's butter, cheese, and poultry became the major source of cash income for the farm families. At the same time it was almost impossible for rural women to support themselves outside male-headed farm households. Investigates rural women's lives in the Nanticoke Valley in New York State. Looks into historical sources, local church records, and dairies and describes women's work on family farms and the patterns of sociability in this area. Recognizes women's actions as feminist; the organizational structure of the Grange gave women a place of their own, and they used their distinct organizational role as a base for integration into a larger group. Grange women's feminist radicalism was rooted in a vision of kinship where the women sought to maintain forms of social relations to which both the value of women's labor and the values of equality and reciprocity were fundamental. Includes a bibliography and an index.

58. Racine, Philip N. "Emily Lyles Harris: A Piedmont Farmer During the Civil War." *South Atlantic Quarterly* 79:4 (1980): 386-397.

Relates records of a wife's entries into a South Carolina farmer's journal while he was in service during the war. Harris made the journal her confidante in which she confided her feelings, fears and opinions. She provided a glimpse of what it might have been like to be a farmer's wife in the middle of the nineteenth century. When war broke out Emily Harris was thirty-three years old, and she had nine children. A set of twins died and the remaining children were between the age of one year and nine months and fourteen years old. In her journal entries, she wrote extensively about the tedious work, the isolation, and her troubled daily life.

59. Rees, Josephine Duggan. "The Women's Land Army." *British Heritage* 12:2 (February 1991): 57-61.

Focuses on the development of the Women's Land Army (W.L.A.) from 1939, when more than 1,000 women were trained and ready for employment at the beginning of World War II. The number of women participating increased by 4,000 a month until 1943, when recruiting stopped, until a total of 80,000 women had joined the ranks. The headquarters was in London and worked in cooperation with the Ministry of Agriculture and Fisheries. Each county had representatives in the headquarters, local representatives who knew each woman and safeguarded her interests and welfare as well as kept in touch with the farmers. The "Land Girls," as they were called, performed a range of agricultural work such as tractor work, hedging, ditching, market gardening, horticulture, and forestry. Includes information about the formation of the Timber Corps in 1942 under the umbrella of W.L.A., when women began to use the great circular saws in sawmill operations.

60. Roberts, Sarah Ellen. *Alberta Homestead: Chronicle of a Pioneer Family.* Austin: University of Texas Press, 1971.

A true story of a family's adventures in homesteading on the prairies of western Canada during the years 1906 to 1912. The mother wrote a daily record of events, and her son, Lathrop Roberts, who prepared the manuscript for publishing, has abridged it somewhat and rearranged some sections to secure better chronological sequence, but he says in the foreword that this is his mother's book. Gives vivid pictures of pioneer life, of hardship and privation, of bitter struggle, years of loneliness, frustration, and disappointment as well as vital and rewarding years.

61. Rose, Margaret. "Traditional and Nontraditional Patterns of Female Activism in the United Farm Workers of America, 1962 to 1980." *Frontiers* 11:1 (1990): 26-32.

Relates the lives and careers of two well known United Farm Worker women, Dolores Huerta, the union cofounder and first vice president, and Helen Chavez, the wife of Cesar Chavez, the United Farm Workers president. Includes an extensive list of references.

62. Sackville-West, V. *The Women's Land Army.* London: Michael Joseph LTD., 1944.

Describes the Land Army as the Cinderella of the women's services. This short book is about the organization. Explains that women were called into service in 1917 because the country had no more than three weeks of food supply left; by the end of the First World War, the land army had 23,000 women enrolled, and during the Second World War, the enrollment figures were 80,000. Women performed all types of farm work including ploughing.

The appendices include statistics on women's employment, the kind of tests they were required to pass, information on the hostels in which they lived, and numerous black and white photos in the appendices.

63.　Schmidt, Elizabeth. "Farmers, Hunters and Gold-Washers: A Reevaluation of Women's Roles in Precolonial and Colonial Zimbabwe." *African Economic History* 17 (1988): 45-80.

Explores the impetus behind the changing gender division of labor and the implications of these alterations for Shona women in the Goromonzi District of Southern Rhodesia (now Zimbabwe) from the late nineteenth century through the 1930s. Examines the diversity of Shona women's roles in the decades prior to colonial rule. Considers gender division of labor in precolonial African societies. While men hunted and young men and boys herded the livestock, the women were primarily responsible for agricultural field work. Gives a realistic view of women's actual participation in a wide range of economic activities, including those frequently categorized as "male" activities. Considers the impact of the decline of African peasantry during the 1920s and 1930s when women started to migrate to farms run by Europeans, to mines and mission stations, and to the towns. Reflects the migration of the African women from the rural areas which had serious implications for the rural African society: female migration was actively discouraged by both European and African male authorities. Describes the way in which changes that took place during the colonial political economy proved to be extremely damaging to the African women. By the 1920s, male household members were increasingly forced to enter the migration labor force. Men's standing within the household increased as their wages increased and surpassed the amount of income from produce, while the position of women in the household declined. Includes numerous references.

64.　Schwieder Dorothy. "Education and Change in the Lives of Iowa Farm Women, 1900-1940." *Agricultural History* 60:2 (1986): 200-215.

Relates the changes that took place for the farm women in the state of Iowa in the first part of the twentieth century: the Department of Agriculture sent out 55,000 letters of inquiry to farm women in 1913 to determine how the department could render better service to them; the department received over 2,000 replies in which farm women commented on many aspects of their lives and offered suggestions for improving rural life. Results highlighted a need for practical home education, and in Iowa the farm women could attend both country and state farm institutes as early as 1900. Within Iowa State College, the Extension Services in Home Economics became the group most closely associated with farm women.

65. Simons, Thordis, ed. *You May Plow Here: The Narrative of Sara Brooks.* New York: Simon & Schuster, Inc., 1986.

Accounts in first person the life of Sara Brooks, who grew up on an Alabama farm in the early 1900s. Recalls in detail the work of cotton picking, harvesting corn, and butchering hogs on the farm. Relates family self-sufficiency in raising all their own food, their hard work and poverty.

66. Shaver, Frances M. "Women, Work and the Evolution of Agriculture." *Journal of Rural Studies* 7:1-2 (1991): 37-43.

Studies data collected during a year of fieldwork in a small Quebec parish which consisted of semi-structured interviews with the adult members of 63 randomly selected farm families. Relies on the clarification of two key elements in the development of agricultural production in the last four decades: the modernization of agriculture and the development of capitalist agriculture. Sheds light on the forces which account for variation in activities and argues the need for an analytical framework sensitive to the complex nature of capitalist development in agriculture. Sets a framework in relation to Canadian agriculture and applies data collected in the farming community in Quebec. Reveals important contribution of women's work in spite of the radical transformation in the agricultural sector. Sees a pattern that is, however, peculiar only to women and states that data presented in another paper show that the situation for men is reversed.

67. Stanley, Autumn. *Mothers and Daughters of Invention: Notes for a Revised History of Technology.* Metuchen, N.J.: The Scarecrow Press, 1993, pp.12-124.

This impressive book of over 1100 pages starts out with two chapters on a historical perspective of horticulture and agriculture, and relates women's inventions in these fields. Covers innovations in irrigation, food processing and cooking, food storage and preservation as well as plant culture, animal husbandry, agricultural machines, and genetic engineering such as Frances Willoughby's 1722 patent of a grain-thrashing machine in England and Lady Ann Vavasour's patent #9405 for a land-tilling machine in 1842 as well as a travel book to raise money for an agricultural school for Irish youth with little or no money. Describes Vavasour's tiller and how she came to invent it. Reports on women working on research teams at the University of Wisconsin, Cornell University and Stanford University, to make a breakthrough that will have worldwide implications for savings of mineral and energy resources and reduction in the cost of producing food. Includes several appendices on the details of agricultural inventions, a useful bibliography, and a general index.

68. Starr, Karen. "Fighting for a Future: Farm Women of the Nonpartisan League." *Minnesota History* 48:6 (1983): 255-262.

Expresses agrarian populist concerns of farmers in the Nonpartisan League (NPL) throughout the United States at the beginning of the twentieth century through regular statewide newspaper feature pages for women readers, which promoted communication of political ideas among women, left a permanent record of their activity, and provided present day historians with a view of the problems of the farm women.

69. Sturgis, Cynthia. "How're You Gonna Keep 'Em Down on the Farm?: Rural Women and the Urban Model in Utah." *Agricultural History* 60:2 (1986): 182-199.

Focuses on the flight from the farm as a growing concern in the early twentieth century, which prompted agricultural educators to promote domestic science education to change the circumstances and nature of rural women's work and elevate homemaking to the status of a profession which required specific skills. Although the Utah farm women desired such improvements, their need may have been less great, due to the unusual settlement pattern known as the "Mormon village," where farm families lived in small communities and traveled out daily to their work in the fields. The Mormon village reduced isolation and provided more opportunities for social interaction. Examines the needs of the Utah women and how home demonstration agents supervised women's activities both through "Home and Community" sections of the Farm Bureau and as leaders of a network of girls' clubs.

70. Webb, Anne B. "Minnesota Women Homesteaders: 1863-1889." *Journal of Social History* 23 (Fall 1989): 115-136.

A study on frontier women farmers based on a quantitative analysis of homestead records of a sample of 259 women who homesteaded in Minnesota for at least one year without a husband and gained title to the land. It is also supplemented by letters and diaries, tax and census data, and other local records. Answers the questions of who farming women were, what farming was like for them, and what contribution their farming made to the society of which they were a part.

71. Zappi, Elda Gentili. *If Eight Hours Seem Too Few: Mobilization of Women Workers in the Italian Rice Fields.* Albany, N.Y.: State University of New York Press, 1991.

Presents the first accurate picture of the thousands of women who weeded the rice fields in Northern Italy during the early part of the nineteenth century. Includes a wide range of issues such as the history of the female labor force, their daily lives, and effort of the Socialist Party to lure women laborers away from the Catholic Church. Covers the history of Italian feminism and the campaign for women suffrage. Describes labor actions of the rice weeders who were among the most radical of all workers in Italy prior to World War I. Includes an extensive bibliography and index.

WOMEN'S ROLE IN AGRICULTURAL ECONOMIC DEVELOPMENT

Ester Boserup's book *Women's Role in Economic Development*[1] is one of the most cited studies on women in agricultural development. As an economist, Boserup analyzed the modernization of agriculture and presented a comprehensive study of women's role in the developmental process. Until now, the specific role of women had been ignored in all studies on development. Boserup pointed out a variety of subjects that are systematically related to the role of women in the economy. Her work suggested this area as one that required further study and stimulated a flood of multidisciplinary research with contributions from sociology, economics, anthropology, ethnology, geography, and history worldwide.

As studies on women's labor force participation increased, it became obvious that there were few statistical studies capturing the degree of women's participation in economic life. Survey work with detailed studies of women's activities and observation of everyday life, made it clear that the existing labor force statistics had a tendency to underestimate women's contribution to production. The literature also showed that development often has had an adverse impact on women: since agriculture was considered men's domain, women were often denied access to production resources and new technology.

In the 1980s, many agencies, both private and public, started funding studies on development projects. This chapter contains all the information that could be located on women's role in economic development. Africa is the center stage for many of these studies, but we also find research conducted in Asia, South and North America and some studies from Europe. The chapter first covers all the larger geographical areas of the world and then lists general studies.

AFRICA

72. Barrett, Hazel and Angela Browne. "Environmental and Economic Sustainability: Women's Horticultural Production in Gambia." *Geography* 76 (July 1991): 241-248.

---

[1]Boserup, Ester. *Women's Role in Economic Development.* London: Allen & Unwin, 1970; New York: St. Martin's Press, 1970.

Uses women's horticultural projects in Gambia to contrast sustainability of different approaches to environmental management and technology. One top-down scheme is studied in depth with detailed analysis of its environmental and economical viability. Shows agricultural practices and irrigation technology associated with this scheme that make it unsustainable when viewed longitudinally. Suggests small-scale projects that make better use of women's farming skills but involve greater use of their time and energy.

Views women in Africa as victims of well-intentioned projects which undermine the resource base upon which their survival depends. The double crisis of economic stagnation and environmental degradation can only be reversed when the international community accepts the need for long-term sustainable development programs. This will only happen when women's environmental knowledge is recognized and harnessed. A location map and a few references are included.

73. Bay, Edna, ed. *Women and Work in Africa.* Boulder, CO: Westview Press, 1982.

A collection of papers from a symposium held in 1979 at the University of Illinois at Urbana-Champaign to promote the field of economic studies on women in Africa and to provide a forum for exchange of ideas among scholars and professionals concerned about women.

Covers topics on women's production outside the development process, economic change and ideological conflicts, differential effects of development policies, and women and work in Africa. Presents 14 recent case studies with relevant references and statistics.

74. Bembridge, T. J. "The Role of Women in Agricultural and Rural Development in Transkei." *Journal of Contemporary African Studies* 7:1-2 (April/October 1988):149-182.

For the large majority of rural women, development has not meant a change for the better, despite the fact that the participation of women in smallholder farming is vital. The author says overlooking this simple fact may thwart efforts to improve food production and social stability in developing countries. Deals with the role of women in rural Xhosa society and reviews personal, socio-economic, and socio-psychological factors to gain insight into the more complex issues raised by the role of women in rural households and in general development process as experienced in Transkei. A survey was conducted in the mixed-farming regions in the Qumbu and Nqamakwe districts as well as at the Qamatra Irrigation Scheme in the St. Mark's district. The sample size varied from 5 to 10% depending on the population figures for the different areas. Includes many statistical tables and an extensive list of references.

75. Boserup, Ester. *Women's Role in Economic Development.* London: Allen & Unwin, 1970; New York: St. Martin's Press, 1970.

One of the most cited studies of women in agricultural development. Ester Boserup, an economist, was one of the first to note the prevalence in Africa of farming systems where all the tasks connected to food production are done by women. Boserup documented the ties between certain land use patterns and agricultural techniques and the division of labor based on gender. She regarded the African as the prototype of female farming system. Analyzes the modernization of agriculture and migration to the cities, and its effect in changing the sexual patterns of productive work. Links these patterns to population density, landholding systems, and technology. Expresses concern that this transition would deprive women of their productive roles in the developing African societies and urged further study of these trends. There are over 60 statistical tables, references, and an index.

76. Bryson, Judith. "Women and Agriculture in Sub-Saharan Africa: Implications for Development (An Exploratory Study)." *Journal of Development Studies* 17:3 (1981): 28-45.

Clarifies the interaction between production systems and social systems in Sub-Saharan Africa and looks at data on the importance of women's role in agriculture. Discusses results with respect to their impact on the two-sector "agriculture-industry" development models and their implications for the future. Expresses view that women's role in agriculture supported past development, but the failure to recognize and enhance women's activities is contributing to current problems with the food supply, and that this could be overcome most effectively by working with rather than against women. References are included.

77. Bukh, Jette. *The Village Women in Ghana*. Uppsala: Scandinavian Institute of African Studies, 1979.

   Presents an analysis of data collected in a village, Tsito, in southeastern Ghana in 1973 and 1976-77. Discusses changes in the roles of both men and women in the public and domestic spheres of agricultural production. Asserts that the development of the market economy and the introduction of cocoa cultivation brought a new social division of labor that allocated to women a major role within subsistence activities.
   Men were drawn into the cash economy, first as cocoa producers and later on, when the conditions for cocoa production changed, as migrant workers. Structural transformation created more freedom for women. However, they were caught between their traditional responsibility for their children and lack of control over necessary economic resources like land, labor, and money. The number of divorces increased in the villages, most often on the initiative of the women. Statistical tables and a bibliography are included.

78. Callaway, Barbara. "The Socialization and Seclusion of Hausa Women." *The Journal of Modern African Studies* 22:3 (1984): 429-450.

   The main part of this study refers to Hausa women in the predominantly Muslim city of Kano, Nigeria. Includes information about women in nearby villages in Zaria and their income-earning occupations such as food processing and preparation, spinning, weaving, and mat making. Unlike elsewhere in West Africa and other areas of Nigeria itself, women in rural Hausa communities are excluded from direct farm operation primarily because of the constraining influence of Islam.

79. Carney, Judith A. "Peasant Women and Economic Transformation in the Gambia." *Development and Change* 23:2 (April 1990): 67-90.

   Discusses gender conflict rife in Gambia irrigation projects. Points to the significance of female labor in relation to contemporary patterns of agrarian transformation as well as linkage between women's access to land for independent farming and forms of project participation. Women were to become the principal beneficiaries of Gambian irrigated rice and vegetable projects and benefit from commercialized production of their traditional crops. The donor's equity concerns were inscribed within a broader policy context intended to restructure and intensify labor routines through contract farming. Includes references.

80. Carr, Marilyn, ed. *Women and Food Security: The Experience of the SADCC Countries.* London: Intermediate Publications, 1991.

The Government of Tanzania hosted a regional conference in Arusha in 1988 to identify food technology projects of benefit to women which could be channeled through the SADCC mechanism. In order to have as much information as possible on women and food technologies, each of the nine member states was asked to prepare a background country paper on women's access to and use of improved food related technologies. The result is a collection of papers about access to credit and training; involvement in technology design and adaptation; availability of raw materials, in infrastructure, and markets; and the political environment. Includes examples of successful projects which provided a framework for discussion of potential replication in other countries in the region. Countries included in the SADCC region: Malawi, Zambia, Botswana, Zimbabwe, Tanzania, Angola, Mozambique, Lesotho, Swaziland.

81. Chapman Smock, Audrey. "Measuring Rural Women's Economic Roles and Contributions in Kenya." *Studies in Family Planning* 10:11-12 (1979): 385-390.

Explains sample survey programs conducted by the Bureau of Statistics in Kenya. The Integrated Rural Survey (IRS) is an annual socioeconomic survey, in which a variety of questionnaires are administered to the same households within the national sample. To complement and supplement existing survey data, the Central Bureau of Statistics developed special modules specifying individual rather than household data. One such survey, the Division of Labour Module, was administered in February and March 1979. Includes some of the questionnaires.

82. Cheater, Angela. "Women and Their Participation in Commercial Agricultural Production: The Case of Medium-Scale Freehold in Zimbabwe." *Development and Change* 12:3 (1981): 349-377.

   Examines data on women's role in commercial agriculture from the author's fieldwork conducted during 1973-74 in an area approximately 60 miles west of the capital, Salisbury, in Zimbabwe. Differentiates the participation of women in commercial agricultural production from their roles in the peasant system, thereby highlighting internal intricacies of behavior, attitudes, conflicts, and contradictions. Argues that "commercial production may rest, not simply on capitalism, but on what initially may appear to be a rather strange mixture of capitalist forces of production and peasant relations of production."

83. Clark, Barbara A. "The Work Done by Rural Women in Malawi." *Eastern African Journal of Rural Development* 8:2 (1975): 80-90.

   Reports on results of a survey of village farm families during 1970-71. Data from five survey sites were used: Mwakasangila and Mwangosi villages in Karonga District; Filimoni Ngulube, Kawinga Banda and Thoza villages in Mzinba District; Khuguwe and Chalingana villages in Thyolo District; Maniwa villages in the Kasupe District: and Chapomoka and Kalikopo villages, which are part of the Chikwawa Cotton Development Project. Includes many statistical tables.

84. Conti, Anna. "Capitalist Organization of Production through Non-Capitalist Relations: Women's Role in a Pilot Resettlement in Upper-Volta." *Review of African Political Economy* 15-16 (1979):75-92.

   Discusses Upper Volta pilot resettlement scheme designed as model for multi-million dollar projects financed by the World Bank and others. Found to be nearly intolerable to women, due to the lack of basic facilities such as market places, adequate land for family food production, village wells, etc. Analyzes the structure of labor power required by the Amenagement des Valles des Volta (AVV) project and identifies two elements: production and reproduction.

85. Creevey, Lucy E. "Supporting Small-Scale Enterprises for Women Farmers in the Sahel." *Journal of International Development* 3:4 (1991): 355-386.

   Examines the reason for success and failures of small-scale enterprise programs for rural Sahel women. The CILCA-Mali program has assisted with projects in rain-fed crop production, vegetable production, sheep fattening, poultry raising, and soap and textile production. Emphasizes training and support for the organization of village women's groups.

86. Creevey, Lucy E., ed. *Women Farmers in Africa: Rural Development in Mali and the Sahel.* N.Y.: Syracuse University Press, 1986.

   The editor is director of the program "Appropriate Technology and Energy Management for Development" and professor of City and Regional Planning at the University of Pennsylvania. Contains papers which were presented at the Bamako Workshop on Training and Animation of Rural Women sponsored by the Food Corps Program International (CILCA) and co-sponsored by the Union of Mali Women. Includes papers on women farmers in Mali and the Sahel, case studies of ongoing projects on development programs for rural women in these areas. Appendix contains facts on the Sahel, an extensive bibliography, and an index.

87. Davison, Jean, ed. *Agriculture, Women and Land: The African Experience.* Boulder, CO: Westview Press, 1988.

   Impact of colonialism and integration into the world economy, together with increasing populations, have weakened women's access to land, labor, and capital while increasing women's responsibilities for agricultural production. Includes papers about selected major regions of African societies where women play a primary role in agricultural production and reflect changing practices in land use and agriculture. Gives an historical overview of the regions covered: West Africa, Central Africa. Kathleen Cloud and Jane Knowles give recommendations for action. Includes references and information about contributors at the end of each chapter.

88.   Dey, Jennie. *Women in Food Production and Food Security in Africa.* Rome: Food and Agriculture Organization of the United Nations, 1984.

Reviews women's responsibilities for cash and staple crop production, for secondary and gathered foods, for animal production, fisheries, and food handling within the context of food security. Proposes measures to increase farm output and efficient food production and suggests governmental and international initiatives to strengthen policies related to women and food security as well as greater access to resources and credit. Bibliography included.

89.   Due, Jean M."Women Made Visible: Their Contributions to Farming Systems and Household Incomes in Zambia and Tanzania." *Culture and Agriculture* 26 (1985): 16-19.

Discusses studies of women's contributions to farming systems and household incomes in Tanzania and Zambia, where females contribute half of the labor for major crops and more than one half the labor for vegetables and fruit production. Relates decision-making role in determining what crops are planted and sold.

90.   Due, Jean M., and Marcia White. "Contrast Between Joint and Female-Headed Farm Households in Zambia." *Eastern Africa Economic Review* 2:1 (1986): 94-98.

Examines differences in household size, acreage in crops and income earned between joint households and female-headed farm households in Zambia. Found that female-headed households plant less acreage of different crops than joint households. With less labor available for crop production, more of their labor was allocated to opportunities for a higher return, such as beer brewing and selling small quantities of fruit and vegetables than their joint household counterparts. Found that female-headed households are poorer and received fewer extension services. Statistics and references are included.

91.   Due, Jean M., et al. "Farming Systems in Three Different Areas of Tanzania 1980-1982." *Zimbabwe Journal of Economics* 1 (1987): 13-21.

Studies similarities and differences in three farming systems in Tanzania as farm families adapt to rainfall, ecology, and soil condition. Women provide more than half the labor for the production of vegetable crops and other major crops, contribute significantly to decisions to plant bean seeds, and are equal recipients of any new high-yielding varieties of beans.

92. Eldredge, Elizabeth A. "Women in Production: The Economic Role of Women in Nineteenth-Century Lesotho." *Signs: Journal of Women in Culture and Society* 16:4 (1991): 707-731.

Points to the general exclusion of women in the study of African economic history and provides a case study of rural production in precolonial Africa that focuses on the role of women. Analyzes contributions to the process of economic expansion in nineteenth-century BaSotho society where women were the motivating force behind agricultural expansion as well as the production of goods and services necessary for the domestic and social system. Demonstrates the role of women as the central force which generated economic change, an essential factor for a clearer understanding of African economic history.

93. Feldman, Rayah. "Women's Groups and Women's Subordination: An Analysis of Policies Towards Rural Women in Kenya." *Review of African Political Economy* 27-28 (1984): 67-85.

In Kenya women already play an integral part in the economy. They are engaged in agricultural labor on their own farmsteads, in wage labor on small farms, in factories, and on large farms and plantations. Women have, however, a clearly defined and subordinated role in all of these areas of work. The Kenyan government made a formal commitment to alleviate their situation. Examines groups coordinated by the Kenyan Women's Bureau which reported sparse impact of uneven and dubious value.

94. Fortmann, Louise. "Economic Status and Women's Participation in Agriculture: A Botswana Case Study." *Rural Sociology* 49:3 (1984): 452-464.

A random sample of 30 households were interviewed in 12 sites, resulting in 358 usable interviews. Follow-up interviews were administered one and four months later, resulting in 358 and 347 interviews respectively. Shows that female farmers differ from male farmers primarily in terms of resources. When access to resources is similar, behavior on the part of farmers tends to be similar, as does their relationship with government staff. Includes references.

95. Gittinger, J. Price, et al. *Household Food Security and the Role of Women.* World Bank Discussion Paper No.96. Washington, D.C.: The World Bank, 1990.

    Reports on the Symposium on Household Food Security and the Role of Women held in Kadoma, Zimbabwe, in January 1990. Forty-seven senior African policymakers, program administrators, academic specialists, and staff of international agencies from seven countries in east and southern Africa focused on constraints that women face and practical measures to reduce them. Issues addressed were nutrition programs for low-income households, women's access to credit, and extension advice and technology. The deliberations led participants to formulate guidelines for African policymakers and donors. Pointed to the importance of increasing the benefits and improving decision-making authority of women. Increasing women's rights would expand opportunities for both men and women. Includes a list of the participants.

96. Gladwin, Christine H., and Della McMillan. "Is a Turnaround in Africa Possible Without Helping African Women To Farm?" *Economic Development and Cultural Change* 37 (January 1989): 345-369.

    This well-researched paper shows that in the short run it is not possible to make a turnaround in African agriculture without helping the women farmers. Urges all policy planners on international, national, and local levels to incorporate women as agricultural producers with full access to yield-increasing inputs in development projects in order to avoid female displacement. Includes statistical tables and references.

97. Gladwin, Christine H., et al. "Providing Africa's Women Farmers Access: One Solution to the Food Crisis." *Journal of African Studies* 13:4 (Winter 1986-87): 131-141.

   Discusses several projects that failed because women were ignored as well as several case studies that were successful because they incorporated women in their projects. Urges policy planners to provide technology, credit advice, and extension services to women farmers who participate actively in agriculture. Recommends a variety of projects aimed at solving the food crisis.

98. Gladwin, Christina H., ed. *Structural Adjustment and African Women Farmers*. Center for African Studies, University of Florida, Gainesville: University of Florida Press, 1991.

   A collection of 16 papers presenting viewpoints from noted African and Africanist social scientists concerning the debate about Structural Adjustment Programs (SAPs). Covers structural adjustment and its impact on women farmers, adjustment policies, gender and the economy, and future prospects.

99. Goody, Jack, and Joan Buckley. "Inheritance and Women's Labour in Africa." *Africa* 73 (1973): 108-121.

   Raises concerns about women's contributions to cultivation in Africa and their lack of overall control, which is largely in the hands of men. Found that their contributions to agricultural production are greater where the basic means of land control is inherited matrilineal. Most African societies that practice hoe culture or female farming are marked by patrilineal inheritance. Questions the influence of this form of division of labor on patrilineal institutions. Found that female farming is strongly linked to two types of inheritance: matrilineal systems and vertical inheritance called "the house-property complex" after Gluckman.

100. Grier, Beverly. "Pawns, Porters and Petty Traders: Women in Transition to Cash Crop Agriculture in Colonial Ghana." *Signs: Journal of Women and Culture in Society* 17:2 (Winter 1992): 304-328.

Focuses on the role of gender relations in the transition to export agriculture in Ghana during the late nineteenth and early twentieth centuries. Looks at the relationship between women's labor and the process of capital accumulation at global and local levels and found that cocoa traders and rural creditors were overwhelmingly men. Examines the mechanism by which the colonial state reinforced a particular role for women in the rural economy. The essay is divided into three parts: a reconstruction of gender and class relations in precolonial southern Ghana, an examination of changes generated by the rapid expansion of cocoa production and the ways in which the continued subordination of women and the exploitation of their labor were central to that expansion, and a review of the colonial state as an important factor that guaranteed a mostly female unpaid agricultural labor force.

101. Guyer, Jane I. "Female Farming and the Evolution of Food Production Patterns among the Beti of South Central Cameroon." *Africa* 50:4 (1980): 341-356.

Through an ethnographic study of food production, this study focuses on the impact of the cash crop economy on farming practices. Argues that both continuities and changes are related to the way in which the Beti division of labor by sex has adjusted to the cash cropping of cocoa. Cocoa cultivation has altered the relationship of population to the land available for food farming, and it has made farmers' income more dependent on market prices than in the past. Looks at the Beti food economy before the penetration of the colonial economy by interviewing elders in six villages representing different ethnic groups. Documents changes which have taken place via published studies and field work data from two Eton villages. Pattern of food production remained within the framework of traditional fields and rotations but shifted in emphasis towards a variant in which female labor alone could produce a nutritionally adequate and culturally acceptable diet. Increased urban demand for food in Southern Cameroon has begun to offer women a way of making a regular cash income as well as an alternative to cocoa production for men.

102. Guyer, Jane I. "The Multiplication of Labor. Historical Methods in the Study of Gender and Agricultural Change in Modern Africa." *Current Anthropology* 29:2 (1988): 247-272.

Outlines the ethnography and history of female farming among the Beti of Southern Cameroon and includes a survey of the main traditions of work on the division of labor within the field of anthropology. Points out the recurrent minor theme of rhythmic structures as a descriptive method amenable to the essential linking of material conditions, social control, and cultural value. The final part casts the Beti case in a new light, showing how a focus on rhythms can provide the empirical basis for tracing historical change in a way which can accommodate what might otherwise be seen as contrasting processes. Many comments by other authors follow the article as well as a reply by Ms. Guyer.

103. Guyer, Jane I. "Women's Work and Production Systems: A Review of Two Reports on the Agricultural Crisis." *Review of African Political Economy* 27-28 (1983): 186-192.

The two reports on the African agricultural crises which are reviewed here are *Accelerated Development in Sub-Saharan Africa: An Agenda for Action* (The World Bank, Washington, D.C., 1981) and *Food Problems and Prospects in Sub-Saharan Africa: The Decade of the 1980s* (United States Department of Agriculture, Washington, D.C., 1981).

104. Hafkin, Nancy J., and Edna G. Bay, eds. *Women in Africa: Studies in Social and Economic Change*. Stanford, CA: Stanford University Press, 1976.

Examines the culmination of a project of the Women's Committee of the African Studies Association to remedy two perceived problems associated with women's issues: the relatively scant literature on African women and the difficulty female scholars have getting their work published. Emphasis is on women and change in Africa, but change in two senses. First, articles analyze women not as objects, but as actors, documenting the wide variety of activities in which African women have participated. Second, this new perspective, in turn, recognizes that women in Africa act as agents of change within their own societies. The collection presents a picture of women who have sought to control their lives and who have understood and dealt with the forces that affect them. It also includes directions that research on women is likely to take. References and an index are included.

105. Hanger, Jane, and Jon Morris. "Women and the Household Economy" in Robert Chambers and Jon Morris, eds. *MWEA, An Irrigated Rice Settlement in Kenya*. New York: Humanities Press, Inc., 1973, pp.209-244.

The data used in this chapter come from a large study by Jane Hanger on the place of women in peasant farming systems. Discusses village life, women's use of services, the role of women in off-scheme farming systems, women's time allocation, and women's responsibilities. Statistical tables and references are included.

106. Haugerud, Angelique. "The Consequences of Land Tenure Reform among Smallholders in the Kenya Highlands." *Rural Africana* 15-16 (Winter-Spring 1983): 65-89.

Argues that Kenyan land tenure reform has not acted as a vehicle for capitalizing agriculture and encouraging agrarian entrepreneurship, but it has created rural inequalities associated with nonfarm income and access to state resources. Documents the extent to which informal social relations contradict the formal system of title ownership and looks briefly at implications of the informal system's declining capacity to absorb the landless and to slow the emergence of a rural proletariat. The study is based on field research in the highlands of Embu District, one of the first districts to undergo land reform just prior to independence.

107. Hay, Margaret Jean, and Sharon Stichter, eds. *African Women South of the Sahara*. London: Longman, Inc., 1984.

This textbook is divided into three parts covering African women in the economy, in the society and culture, and in politics and policy. The appendix contains selected statistics on African women, a bibliography, background information on the contributors, and an index.

108. Henn, Jeanne Koopman. "Feeding the Cities and Feeding the Peasants: What Role for Africa's Women Farmers?" *World Development* 11:12 (1983): 1043-1051.

The extensive farming system of the Beti people of southern

Cameroon and the women farmers of the Haya of Northwestern Tanzania is examined in this study. The author looks at the technological and socioeconomic problems constraining the expansion of food production and marketing. Asserts that a policy to develop cooperative village farms to replace individual household production would have a progressive impact on transforming the traditional social relations of rural production and expand agricultural development.

109. Herz, Barbara. "Women in Development: Kenya's Experience." *Finance and Development* 26 (June 1989): 43-45.

Even where national policies encourage equality of opportunity, women generally lag behind men in educational attainment, earning capacity, and many other aspects. Kenya's experience suggests that some of these disadvantages can be overcome at modest cost. The government has established economic policies, development programs, and a legal framework to strengthen incentives and productive capacity for women. Investments in women's education, health care, including family planning, and agriculture extension directly benefit women and can have a strong impact on economic performance, family health, and population growth. Kenya has also started to improve access for women to productive resources, particularly water, fuelwood, and credit, and today more than 60% of women living outside Nairobi are literate.

110. Heyer, Judith. "The Origins of Regional Inequalities in Smallholder Agriculture in Kenya 1920-1973." *Eastern Africa Journal of Rural Development* 8:1-2 (1975): 142-181.

Summarizes briefly the origins of the present disparities in agricultural development in small farm areas in Kenya. Historical material gives wealth of information, but there are also many data gaps. The author researched more than the obvious primary sources using annual reports of the Native Affairs Department, the Agricultural Department, and the Veterinary Department. Points out how behind the impressive growth of the small farm sector, there emerges a picture of substantial inequality as well as an indication of other areas that need to be studied. Many statistical tables are included.

111. Hill, Polly. "Food Farming and Migration from Fante Villages." *Africa* 48:3 (1978): 220-230.

> The author visited several Fante food-growing villages near Cape-Coast in southern Ghana to study the organization of food farming with special reference to the position of women farmers. However, much of her time was spent on an investigation of the outward migration of nearly all young men and women in the area. Found a high incidence of spouselessness among men as well as women.

112. Hirchmann, David, and Megan Vaugh. "Food Production and Income Generation in Matrilineal Society: Rural Women in Zomba, Malawi." *Journal of Southern African Studies* 10:1 (1983): 86-99.

> Describes the various food-producing and income-generating patterns found in rural households in the Zomba district of Malawi. Focuses on women but also links women closely to their own household and the wider family groupings. Seventy women were interviewed, and 57% of them said they did not have enough land to grow adequate supply of staple food to feed their families. The main crop is maize and cassava, but a majority of the women interviewed grew their maize mixed with other crops. There is a high divorce rate, and about one third of the women interviewed were unmarried, divorced, permanently separated or widowed, and 14 % had husbands who lived most of the time elsewhere.

113. Hurtado, Magdalena A., and Kim Hill. "Experimental Studies of Tool Efficiency Among Machiguenga Women and Implications for Root-Digging Foragers." *Journal of Anthropological Research* 45 (Summer 1989): 207-217.

> Efficiency of traditional wooden and modern metal tools were measured among the Machiguenga women of the Manu River Basin. Findings indicate that individuals using the wooden tools used two to three times as many minutes digging and peeling one kilogram of manioca than those using machetes and knives. It shows, however, that the adoption of metal tools is only likely to lead to important changes in diet and time allocation when the return rates in root digging and processing are

significantly lower than the ones found among Machiguenga women. The results should, therefore, make us cautious about assuming that adoption of modern technology severely alters native lifestyles in ways that cannot be predicted or understood. Statistical tables and a few references are included.

114. ILO Tripartite African Regional Seminar on Rural Development and Women, Dakar, Senegal. *Rural Development and Women in Africa.* Geneva: International Labour Office, 1984.

   Proceedings of a seminar organized by the World Employment Program and the Association of African Women for Research and Development that took place in Dakar, Senegal, in 1981. Purpose of the seminar was to debate priority issues and recommend guidelines for policy-makers and planners for the improvement of the living and working conditions of rural women in Africa. The first part of the book includes an overview of the seminar, and the second part contains the 11 papers presented. References are included.

115. Jarosz, Lucy Antonia. *The "Traffic in Women": Buying and Selling Labor Power in African Contract Farming.* M.A. Thesis, University of Berkeley, 1987.

   Explores contract farming's role in the process of agrarian change in Africa. Contract farmers commit to supply agricultural products to specifications set out in advance by an oral or written contract. This way of farming introduces substantial changes in peasant household organization, and the author explores peasant theory regarding household organization and labor process in light of empirical data from the contract production of tobacco, sugar, and vegetables in Kenya.

116. Jarosz, Lucy. "Women as Rice Sharecroppers in Madagascar." *Society and Natural Resources* 4:1 (1991): 53-63.

   Presents a case study drawn from a rice-growing region in Madagascar and examines the roles of gender and class in regional rice production in order to demonstrate that sharecropping is one strategy male and female households heads use to gain access to resources, and, in so doing, reinforce relations structured by kinship, gender, and class. In her study Jarosz found that sharecropping did not uniformly impede or facilitate a transition to agrarian capitalism. Rather, its effects were uneven and ambiguous, increasing productivity for some and leading to economic stagnation and dispossession for others.

117. Kaberry, Phyllis M. *Women of the Grassfields: A Study of the Economic Position of Women in Bamenda, British Cameroons.* London: Her Majesty's Stationary Office, 1952.

Analyzes the role women play in agriculture economics, the system of land tenure, pastoral and trade development. Looks also at the ecological conditions, the type of natural resources, the traditional means of exploiting them, the range of economic needs, and the bearing of all such factors on the division of labor between the sexes. Fifteen months were spent in Kimbaw and other villages of Nsaw to gather data. Statistical data and detailed economic information from 16 different households are included in the appendix as well as sketch maps and an index.

118. Koenig, Dolores. "Alternative Views of 'The Energy Problem': Why Malian Villages Have Other Priorities." *Human Organization* 45 (1986): 170-176.

This study was part of a larger study to support the solar energy laboratory in Bamako. The emphasis is on women, because it is they who do most of the major activities discussed such as collecting the fuel and preparing and cooking the food. Covers four zones, each including five villages, and efforts were made to get reliable information about energy and time use. The data showed that the search for fuel is a relatively small part of the cooking process. It also showed that the average woman spent between one and one half to two hours per day in cooking and preparing one meal; that meant up to seven hours per day for three meals. However, the women rotated the cooking tasks among the wives of the household; each woman cooked one day for the family, thus the women had an advantage when there was more than one adult woman in the family. The study showed that the highest priority for villagers was more reliable and more accessible water supplies. The regular search for water is one of the more onerous activities that women perform. The paper presents an alternate way of looking at the energy problem with the hope that this may lead to a development strategy with different priorities for intervention and be more in accord with the needs of the local population.

119. Kossoudji, Sherrie, and Eva Mueller. "The Economic and Demographic Status of Female-Headed Households in Rural Botswana." *Economic Development and Cultural Change* 31:4 (1983): 831-859.

An analysis of the Rural Income Distribution Survey (RIDS) conducted in 1974-75 by the Central Statistical Office of Botswana with financial support from the World Bank. The survey covered 957 households containing 6,475 individuals, and its primary objective was to collect data on income distribution in rural areas. Each household was visited once a month over a period of 12 months. The authors list several strategies to improve the income of women more directly and immediately. Female-headed households seem to be as competent in crop production as male-headed households, and the authors point out that instruction on more efficient means of cultivation and the use of productive inputs should be directed at both women and men. Statistical tables and references are included.

120. Lefaucheux, Marie-Helen. "The Contribution of Women to the Economic Development and Social Development of African Countries." *International Labour Review* 66:1 (1962): 15-30.

The author, the president of the International Council of Women, discusses the subject of equality, particularly in employment. The methods of obtaining equality may differ between countries and between areas of the globe, but in the developing countries in Africa, the best way to ensure that women have their rightful place is to find ways of including them in the very process of economic development. African women now stand on the threshold of many new openings for activity in a wide range of spheres.

121. LeVine, Robert A. "Sex Roles and Economic Change in Africa." *Ethnology* 4:2 (April 1966): 186-193.

In the agricultural societies of sub-Saharan Africa the author found uniformity in the traditional division of labor by sex and the husband-wife relationship. The husbands clear the bush, but the wives have the larger share of the routine cultivation. The women contribute very heavily to the base economy, while the men have the more prestigious activities such as cattle ranching, government, and legislation. However, in the Nigerian societies economic development has, through expansion of the traditional marketing role, allowed wives to attain independent and sometimes greater incomes than their husbands, and male domination in marital relations has been seriously challenged. The general validity of this analysis remains to be tested in systematic research.

122. Martin, Susan. "Gender and Innovation: Farming, Cooking and Palm Processing in the Ngwa Region of Southeastern Nigeria 1900-1930." *Journal of African History* 25 (1984): 411-427.

Between 1900 and 1930 women did most of the work in food farming and cash crop production in the Ngwa region in Nigeria. The scarcity of female labor was the main factor which restricted the expansion of palm production and stimulated labor-saving innovation. Three main forms of agrarian change are shown to have occurred in the region during this period. Argues that gender relations within the Ngwa society had a strong influence on local patterns of labor scarcity and on the character of the innovations adopted to save labor. In fact, the divisions within Ngwa society were as important as local environmental factors or market forces in shaping the pattern of local agrarian change.

123. Mbata, J.N. and C.T. Amadi. "The Role of Women in Traditional Agriculture: A Case Study of Women in Food Crop Production in River State, Nigeria." *The Ahfad Journal* 7:1 (1990): 32-50.

Over the past decade agricultural productivity has declined, and domestic food production has failed to satisfy local demand in Nigeria. This study investigates the role of women's contribution in food crop production in River State Nigeria. Findings show that women are actively engaged and even dominate many food crop production stages. It was also observed that women overutilize their family labor input, while hired labor, farm size, and seed input were underutilized. In order to attain economic optimum, women should increase their use of hired labor, farm size, and seed inputs, and they should decrease family labor input to the point where the marginal value product of each input is equal to the cost of acquisition of that particular input. Statistical tables and references are included.

124. Mbilinyi, Majorie. "Agribusiness and Women Peasants in Tanzania." *Development and Change* 19:4 (October 1988): 549-583.

The author challenges the analysis of agriculture and the agricultural crisis in Tanzania, which is presented by the World Bank and mainstream social science. According to the mainstream view, Tanzania is a peasant economy and Tanzanian agriculture is based upon peasant household production, which has experienced a steady decline in output and productivity.
Gives a brief history of agriculture in Tanzania, and discusses the feminization of waged and unwaged agricultural labor, and polarization and proletarianization in Rungwe. Shows how smallholder programs as developed by the World Bank have led to disintegration of these holdings and increased women's resistance. The author also provides an alternative analysis of the organization of capital and labor in agriculture, as illustrated by the case study on relations in tea production in Rungwe.

125. Mbilinyi, Majorie J. "The State of Women in Tanzania: Work by Everyone and Exploitation by None?" *Canadian Journal of African Studies* 6:2 (1972): 371-377.

Analyzes the traditional place of women in the Tanzanian African society and the contemporary changes taken place. There are 3.3 million adult women living in the rural areas. The dual role of wife and mother is traditionally the most important for a woman, and children are valued as economic assets. In the subsistence rural economy, children provide important labor resources, and parents expect their children, particularly the boys, to provide for them in old age, while the girls are expected to bring wealth into the family through their brideprice. Modernization has had mixed effects on the women's "world," because many of the ritual and community factors are disappearing. Women's work has increased and unequal educational opportunity still exists. The author sees the new Marriage Contract Law as an attempt by the government to offset some of the inequities faced by women, and she finds the effort of collective production in agriculture as the most exciting development.

126. McHenry, Dean E., Jr. "Communal Farming in Tanzania: A Comparison of Male and Female Participants." *African Studies Review* 25:4 (December 1982): 49-64.

Studies the confusion that persists among villagers over the meaning of the concept communal farming and how men and women perceive it differently. However, the data found that differences do exist between male and female participants in communal work in the degree of their participation and in the effects of their characteristics and attitudes on their degree of participation, but the differences are not substantial. The differences looked at are wealth, age, education, party membership, satisfaction with life, aspiration for the good life, motives for joining the village, like/dislike of communal work. Statistical tables and references are included.

127. Mikell, Gwendolyn. "Filiation, Economic Crisis, and the Status of Women in Rural Ghana." *Canadian Journal of African Studies* 18:1 (1984): 195-218.

Based on data collected in the Sunyani District of Ghana in 1972, 1973, and 1981, where interviews were conducted with 232 cocoa farmers, the analysis provides a context for understanding the Brong matrilineal base and the socio-economic behavior of Brong male and female cocoa farmers. The author provides a historical perspective of the Akan lineage principles and discusses the data gathered. The dynamics of the capitalist system in combination with structural principles within lineage systems tend to produce a situation which is detrimental to women's economic autonomy at points of economic crisis or resource restriction.

128. Mikell, Gwendolyn. "Ghanaian Females, Rural Economy, and National Stability." *African Studies Review* 29:3 (1986): 67-88.

Uses historical, ethnographic case study, statistical, and political data from cocoa farming areas in Ghana to examine the changing relationship of rural women to economic and national stability. Argues that in recent difficult political and economic climates in some African countries, pressures exerted on rural areas have contributed to a heavy reliance upon female producers. Over time these pressures further contribute to an unstable rural economy, because the exploitation of the female labor force, while itself a reaction to socioeconomic trauma, further discourages male involvement in agricultural production. Some long-range suggestions for the future might include the development of female-run agriculture projects on local government land, using subsidized inputs or voluntary labor. References are included.

129. Mkandawire, Richard M. "Customary Land, the State and Agrarian Change in Malawi: The Case of the Chewa Peasantry in the Lilongwe Rural Development Project." *Journal of Contemporary African Studies* 3:1-2 (1984): 109-128.

In 1969 the Malawi government introduced a new land law which was intended to stimulate the country's agrarian development. The Lilongwe Rural Development Project became the first area where the new land law was applied. This article focuses on some of the consequences and long-term implications of the new land concept among the Chewa peasantry in this project. It also questions if the new system will enable the peasants to use their land more productively. The author argues that the new system may only give rise to conflicts between officials and the peasantry but also among the peasants themselves. Traditionally among the Chewa individual households have total control over plots of land they occupy.

130. Mkandawire, Richard M. "Invisible Farmers: Women in Agriculture in South Africa. A Case Study of Malawi." *Journal of Extension Systems* 5:1 (1989): 23-32.

Using Malawi as a case study the author found that 28.8% of rural households are headed by women, and 51% of decisions related to agriculture production are made exclusively by women. However, most programs of planned change are directed at men rather than women, mainly because women have not yet been recognized as primary cultivators. The study showed that 64.5% of women reported they had absolutely no technological advice from rural development extension agents. The author feels that these findings demand re-examination of development strategies and a search for alternatives which could better help rural women.

131. Monson, Jamie, and Marion Kalb, eds. *Women as Food Producers in Developing Countries.* Los Angeles, CA: UCLA African Studies Center, 1985.

  Papers presented at a conference in March 1984 organized by the UCLA African Studies Center and the Los Angeles Office of OEF International, and many other community and university programs. The conference centered on four basic goals: (1) to learn more about the important role played by women throughout the world as food providers, (2) to gain an understanding of the barriers women face as they struggle to feed their families, (3) to find out how these barriers can be overcome, and (4) to investigate ways in which communities can be involved to help.
  The papers are: "Women, Population, and Food: An Overview of the Issue" by Jane Jaquette. "Women's Productivity in Agricultural Households: How Can We Think About It? What Do We Know?" by Kathleen Cloud. "Seeing the Invisible Women Farmers in Africa: Improving Research and Data Collection Methods" by Ruth B. Dixon. "African Women as Small-Scale Entrepreneurs: Their Impact on Employment Creation" by Sylvia White. "Women and the Latin American Livestock Sector" by Suzanna B. Hecht. "Women's Political Consciousness in Africa: A Framework for Analysis" by Kathleen Staudt. "'A Women Has a Voice: Theater for Education and Development in Africa." by Nkeonye Nwankwo. References are included.

132. Murray, Collin. "Marital Strategy in Lesotho: The Redistribution of Migrant Earnings." *African Studies* 35:2 (1976):99-121.

  Examines the intricacies of marriage payments and the migrant labor system, which is the means by which Basotho migrants find cash to establish legitimate marital relationships. The irony is that the migrant labor system is in itself the largest threat to marital stability by enforcing the separation of man and wife for repetitive periods of indefinite duration.

133. Nelson, Nici, ed. *African Women in the Development Process.* London; Totowa, N.J.: Frank Cass and Company Ltd., 1981.

  Information in this book was also published in the *Journal of*

Development Studies 17:3 (April 1981), which is devoted to women's role in the development process. Most of the articles relate to Sub-Saharan Africa and deal with specific issues such as the impact of labor migration on the lives of women, women's cooperatives, and strategies for mobilizing village women.

134. Okelo, Mary. "The Women's Viewpoint" in Obasanjo, Olusegun and Hans D'Orville, eds. *The Challenges of Agricultural Production and Food Security in Africa.* Washington, D.C.: Taylor and Francis, 1992, pp. 83-87.

African women produce 60-80% of the food grown and constitute 46% of the labor force. However women continue to be excluded from the decision-making process, and women's economic contributions are not recorded or reflected in national statistics. In this regard the African Development Bank is taking definite actions to support the role of women in food production.

135. Olenja, Joyce M. "Gender and Agricultural Production in Samia, Kenya: Strategies and Constraints." *Journal of Asian and African Studies* 26:(July-October 1991): 267-275.

Explores the changing roles of women in agricultural production in a patrilineal community of the Samia in Western Kenya. Farm activities are considered demeaning for men; they are expected to work within the public domain. It is, therefore, only women and children who occupy themselves with agricultural work in the domestic domain. Modern formal education takes children away from home, and the women have a substantial burden of sustaining the household. Statistical tables are included.

136. Olmstead, Judith. "Farmer's Wife, Weaver's Wife: Women and Work in Two Southern Ethiopian Communities." *African Studies Review* 18:3 (December 1975): 85-96.

Presents a comparison of women's economic positions in two ethnically similar and economically different communities in the Gamu highlands of Southern Ethiopia. One community engaged both men and women in farming, and only a fraction of their produce was sold. In the other community men were primarily weavers and they migrated to urban centers, while most of the women remained at home where they engaged in agriculture and childcare, and sold some of their produce on the local markets. Statistical tables are included.

137. Orvis, Stephen. *A Patriarchy Transformed: Reproducing Labor and the Viability of Smallholder Agriculture in Kisii.* Nairobi, Kenya: Institute for Development Studies, 1985.

   Presents a historical analysis of the transformation of agricultural production and rural society in Kisii from the precolonial area to the present. Outlines the key elements of the pre-colonial economy, analyzes the colonial transformation of that economy, and provides an analysis of current smallholder economy and its integration with the larger Kenya economy. Addresses the question of how the reproduction of labor takes place and the transformation it has undergone over the last century. The data include historical and anthropological literature on Kisii and eight months of field research.

138. Pala, Achola O. "Toward Models of Development. Definitions of Women and Development: An African Perspective." *Signs: Journal of Women in Culture and Society* 3:1 (1977): 9-13.

   The author wishes to draw attention to points which she considers central to the contemporary position of women in Africa. She suggests that the problems facing African women today, irrespective of their national and social class association, are inextricably bound up in the wider struggle by African people to free themselves from poverty and ideological domination in both intra and international spheres. The question of autonomy and self-determination still remain critical to an understanding of the problems surrounding female participation in contemporary Africa.

139. Rakodi, Carole. "Urban Agriculture: Research Questions and Zambian Evidence." *The Journal of Modern African Studies* 26:3 (1988): 495-515.

   The household economy seems to be neglected in research and policy despite its importance both in adapting to economic change and in survival strategies. The author gives an overview of the economic activities of women in Zambia and discusses land-use planning and policy implications of urban agriculture with the evidence from Zambia. Although it is clear that home-grown food provides a vital or useful supplement for many families, there is little information available of the potential for increasing output.

140. Roberts, Penelope A. "Rural Women's Access to Labor in West Africa" in Sticher, Sharon B., and Jane L. Parpart, eds. *Patriarchy and Class, African Women in the Home and the Workforce.* Boulder: Westview Press, 1988, pp. 97-114.

Explains the complicated hierarchies of gender, rank, generation, and class within and between households in West Africa and points out how women are profoundly disadvantaged within these hierarchies. This has significant consequences for the way in which women enter into production. Examines also the historical circumstances in which women have been able to mobilize labor and clarifies the contemporary constraints on the vast majority who do not.

141. Saito, Katrine A. "Extending Help to Women Farmers in LDSs: What Works and Why." *Finance and Development* 28: (September 1991): 29-31.

In Sub-Saharan Africa, women account for at least 70% of the food staple production, besides participating in food processing and marketing, cash cropping, and animal husbandry. Moreover, increasing numbers of women are becoming heads of households, managing farms on a day to day basis as more men are migrating to the cities. If these women are to carry out their multifaceted roles, they need effective agricultural extension services. Yet the evidence clearly shows that, despite a growing awareness of the need to reach women farmers, these services are generally geared toward male farmers.
Troubled by this highly inefficient use of resources, many policy makers or donors, such as the World Bank, which has invested $2 billion in agricultural extension in 79 countries since the mid-1960s, have begun to work with countries on innovative approaches. Pilot programs are now underway on how to integrate women in their agricultural extension systems. A crucial link in the extension services is making sure that the client is able

to receive the information. One way is through mobile training courses, but the mass media hold the greatest potential.

142. Saito, Katrine A., et al. *Raising the Productivity of Women Farmers in Sub-Saharan Africa.* World Bank Discussion Paper no. 230, Washington, D.C.: The World Bank, 1994.

Presents the findings of the World Bank's executed projects based primarily on four country studies: Burkino Faso, Kenya, Nigeria, and Zambia. Documents women's role in agriculture, identifies, and evaluates the key constraints women face in attempting to raise their productivity, and recommends measures to relieve these constraints. Women's access to agricultural inputs and support services has not improved commensurate with their role as farmers, resulting in considerable loss in agricultural activity and output; more than 20 % according to the analysis from the Kenya study. The report emphasizes on the fact that agricultural development strategies have not adequately focused on the client, who, in Sub-Saharan countries at least, are increasingly women. Statistical tables and references are included.

143. Shapiro, David. "Farm Size, Household Size and Composition, Rural Women's Contribution to Agricultural Production: Evidence from Zaire." *The Journal of Development Studies* 27:1 (October 1990): 1-21.

The data are the outcome of a survey in 1985-1986 of 240 traditional farm households located in the southern part of Zaire, focusing specifically on the questions put forward by H. P. Binswanger and J. McIntire regarding land-abundant tropical agriculture. The author found support for their point concerning self-cultivation, the limitation of hiring of labor, and the importance of the household's age and sex composition in relation to farm-size, and he found that women's contributions were greater than men's. Among the farms included in the data set there is a systematic decline in cultivated area per working household member as the number of working household members increase.

144. Smith, Charles D., and Lesley Stevens. "Farming and Income-Generation in the Female-Headed Smallholder Household: The Case of a Haya Village in Tanzania." *Canadian Journal of African Studies* 22:3 (1988): 552-566.

Compares female- and male-headed farming households in a traditional village located in Muleba District of the West Lake region of Tanzania. Fifty households were interviewed concerning landholdings, crops and yields, inputs and methods of cultivation, division of labor, source of income, household expenses, and health and nutrition. Findings indicate that female-headed smallholder households appear to have distinctive characteristics which dramatically affect their contribution to aggregate food and cash crop production and resemble that of poor male-headed smallholder households, but they lack all access to inherited land under customary law. The author suggests the need for further investigation of this topic.

145. Spring, Anita. "Men and Women Smallholder Participants in a Stall-Feeder Program in Malawi." *Human Organization* 45:2 (1986): 154-162.

Examines the Animal Husbandry Section of the Lilongwe Rural Development Project which started in 1968 and considers how the addition of a livestock enterprise affects the farming systems and farmers' remuneration. Looks at the work patterns and new opportunities which become available for men and women as a result of planned livestock enterprise changes and the recruitment of men and women farmers to the program as well as differences in training, labor, and the use of income gained. Women's involvement in the care of large animals has increased significantly, but their participation is unrecognized by many development staff and consultants and, consequently, few livestock services are directed to women.

146. Spring, Anita. "Putting Women in the Development Agenda: Agricultural Development in Malawi" in Brokenska, David W., and Peter D. Little, eds. *Anthropology of Development and Change in East Africa.* Boulder: Westview Press, 1988, pp.13-42.

This study funded by the Office of Women in Development, United States Agency for International Development (USAID) took place from 1981 to 1983 and was evaluated as the best project that had been carried out by USAID. The director, an anthropologist, explains the project methodology, project design, and the structure of the Agricultural Development Divisions in Malawi.

147. Spring, Anita. *Using Male Research and Extension Personnel to Target Women Farmers.* East Lansing, Michigan: Women in International Development, Michigan State University, Working Paper No. 144, 1985.

Farming Systems and Extension (FSR/E) methodology have several phases (diagnostic, technology design, testing and dissemination) that should include information about sexual division of labor, resource allocation, income generation, and knowledge of farming practice; yet gender is often left out by both researchers.

A case study from Malawi shows that it was not common for women to be included in FSR/E work as trial farmers or in recommendation domains. Subsequently, an extension circular suggested techniques for working with women. Statistical tables and references are included.

148. Spring, Anita. *Women Farmers and Food Issues in Africa: Some Consideration and Suggested Solutions.* East Lansing, Michigan: Women in International Development, Michigan State University, Working Paper No. 139, 1987.

Reviews briefly some of the major aspects of African women's involvement with food production and offers some suggestions for improving participation of women, especially in smallholder agriculture. A chapter on methods for improving extension services to women, statistical data, and references are included.

149. Staudt, Kathleen A. "The Umoja Federation: Women's Cooption into a Local Power Structure." *Western Political Quarterly* 33:2 (1980): 278-290.

Umoja was a women's federation of the late colonial era in western Kenya. Umoja's origins were tied to a political machine headed by an autocratic chief with pressing administrative responsibilities. The style of British colonial "indirect rule" in western Kenya was decentralized from the central perspective but highly centralized at the local level. The author explains the origins of Umoja and the leaders, structures and relations with the local administration, and the organizational transformation and the reasons for Umoja's collapse.

150. Staudt, Kathleen. "Uncaptured or Unmotivated? Women and the Food Crisis in Africa." *Rural Sociology* 52:1 (1987): 37-55.

Reviews a gender approach to agriculture, focusing on the implication of labor differentiation, incentives, and struggles over access to resources for agricultural development. The author then analyzes two approaches to understanding Africa's development crisis: first, the faulty incentives created in statistical strategies, and second the uncaptured peasantry. She concludes the paper with some policy implications.

151. Staudt, Kathleen A. "Women Farmers and Inequities in Agricultural Services." *Rural Africana* 29 (Winter 1975-76): 81-94.

Provides empirical support for the hypothesis that the government gives preference to men in agricultural services. The author proceeds to examine a number of factors that could possibly account for such discrimination. The data were collected in 1975 from 212 small-scale farm households in an administrative location in Kakamega District of western Kenya. The agricultural services in Kenya are described: farm training, loan information, and acquisition. The agricultural instructor visited locations not included in the survey. She found that women farm managers experienced a persistent and pervasive bias in the delivery of agricultural services and that discrimination appears to be the result of prejudice against women.

152. Swindell, Ken. "Family Farms and Migrant Labour: The Strange Farmers of the Gambia." *Canadian Journal of African Studies* 12:1(1978): 3-17.

The Strange Farmers System rests on a host-client basis, whereby a local farmer takes in a migrant on the understanding that he will work for him for between two and four days per week. The stranger is given a plot of land on which he works the rest of the time, cultivating groundnuts which he will sell at the end of the season. The farmer provides the stranger with food and tools, together with a hut within the family compound. The Strange Farmers usually appear in the villages during April and depart in December after the trade season has begun. The author looks at the groundnut cultivation and the role of the Strange Farmers in Gambia, the types of non-family labor input, the farming units using these farmers, and the contribution they make to the groundnut production in Gambia. Found that the presence of the Strange

Farmers makes a material difference to both the total amount of land under cultivation and the amount of groundnut cultivated.

153. Tripp, Robert B. "Time Allocation in Northern Ghana: An Example of the Random Visit Method." *Journal of Developing Areas* 16 (April 1982): 391-400.

Data were collected as part of a study carried out in a Nankane-speaking settlement in Navrongo District in northern Ghana between October 1975 and May 1977. The study uses the technique of random visits, where the researcher selects a sample of households for observation throughout the year and prepares a random schedule of days and hours for visits. The researcher records the activities of all members of the household at the moment of the visit. Over the course of the year, a large series of random observations of community members' activities is compiled, and estimates of the proportion of time spent in various activities can be made. The sample population consisted of 20 adult males, 29 adult females, 11 youths, 15 children, and 12 infants. Only data on the adults' activities are reported here. Several statistical tables are included listing the results of the survey.

154. Tshatsinde, M. A. "Factors Leading to Low Productivity among Rural Women in Agricultural Production." *Agrekon* 29:3 (December 1990): 359-362.

Looks at some of the problems women farmers in South Africa experience which lead to their low productivity. Examines in a sample of 61 female farmers their age, marital status, level of education, number of children, shortage of land, decision making, and marketing. The author found that the majority of female farmers are married and on an average have five children. Their husbands work in urban areas, and the women have little education and have no say in the decision making.

155. Van Allen, Judith. "Women in Africa: Modernization Means More Dependency." *The Center Magazine* 12:3 (May/June 1974): 60-67.

Discusses the facts that Sub-Saharan Africa has always been a "female farming" area and that women are shut out of agricultural training programs, while men are trained to use the newer methods and produce cash crops. The roles into which African women are moving bring benefits, but fewer to them than to men.

156. Vellenga, Dorothy Dee. "Differentiation among Women Farmers in Two Rural Areas in Ghana." *Labour and Society* 2:2 (April 1977): 197-208.

Examines the factors that enable some women to accumulate farms and laborers and restrict other women to food farming and work on cocoa farms belonging to others. The study takes place in the Region of Brong-Ahafo in Ghana. One hundred women were interviewed, and they represent about 10% of the female farmers in both towns. The author was interested in the number of cocoa and food farms the women had, how they had acquired them, whether they worked for anyone else as well as some background information on their age, residence, and family characteristics.

157. Walker, Tjip S. *Innovative Agricultural Extension for Women: A Case Study in Cameroon*. Washington D.C.: World Bank Working Paper, 1990.

Agricultural extension has not been particularly kind to Africa's women farmers; women are under-represented within extension services and benefit much less from improved farming techniques. In Cameroon's Northwest Province, agricultural extension was extended to women in a sustainable, replicable experiment that increased production and women's income. Field work for this report was conducted in August 1988 and included interviews with staff on the national, provincial, and local levels. The success of this program provides a much needed example that extension can be made to work with women using a little creativity and not necessarily at great cost. List of references is provided.

158. Weil, Peter M. "Wet Rice, Women, and Adaptation in Gambia." *Rural Africana* 29 (1973): 20-29.

Examines the political and economic dynamics of a shift in food production among the Mandinka of Gambia, where the production of sorghum by men is being replaced by rice production by women. Demonstrates that this shift is an adaptation to an increasingly commercial economy and is taking place by channeling competition for two vital but scarce resources (tidal swamp lands, and skilled female laborers) through endogenous political and economic mechanisms. The shift in food production is changing the Mandinka economic base while playing a fundamental role in the integration of Mandinka political communities.

159. de Wild, John Charles, ed. *Experiences with Agricultural Development in Tropical Africa.* International Bank for Reconstruction and Development, Baltimore: Johns Hopkins Press, 1967, 2 volumes.

A series of studies on the tropical regions of Africa examines a range of ecological conditions, different approaches to agricultural development, and cultural patterns. Volume 1 is a synthesis of the project and volume 2 presents case studies. The purpose of the project was to evaluate the programs and identify factors which accounted for success and failure. The factors examined included receptivity and incentives to change, labor tenure and land use priorities, agricultural extension, training and education, credit, marketing, and cooperatives.

160. Wipper, Audrey, ed. "The Roles of African Women: Past, Present and Future." *Canadian Journal of African Studies* 6:2. Ottawa: Canadian Association of African Studies, 1972.

This issue of *Canadian Journal of African Studies* brings together articles about women in east, central and west African countries which examine political roles and tactics, marital and family patterns, economic activities, education, and the difficulties in changing women's roles.
African women have had very little participation in the development plans in their respective countries. Policymakers have generally been coalitions of males from Western and African countries who made decisions regarding the roles of women. Three of the articles are in French.

161. Wipper, Audrey, Ed. *Rural Women: Development or Underdevelopment.* East Lansing, Michigan: African Studies Center, Michigan State University, Series: Rural Africana, No. 29, 1976.

Rural Africana, a research bulletin of the African Studies Center at Michigan State University, is devoted to current research in the social sciences, and this issue explores the problems of social and economic development in rural Africa south of the Sahara.

The papers presented look at the effects of social change on women, and the editor stresses the importance of examining the underside of development in three areas: (1) not the formal but the informal sector of the labor force, (2) not the cash crop but the subsistence economy, (3) not the public but the private spheres. The author says, "only when these areas have been explored more fully, can we begin to grasp the breath and depth of women's contribution to development." A selected bibliography on women in Africa is included.

THE AMERICAS

162. Arizpe, Lourdes, and Josefina Aranda. "The 'Comparative Advantage' of Women's Disadvantage: Women Workers in the Strawberry Export Agribusiness in Mexico." *Signs: Journal of Women in Culture and Society* 7:2 (Winter 1981): 453-473.

For several decades, many of the labor-intensive agricultural activities in which women worked as wage laborers have been shifting to developing regions. For example, many jobs formerly held by women in the southern rural areas of the United States have moved south to Mexico and to other Latin American and Caribbean countries. Behind this movement lie both market pressures and the rationale of "comparative advantage," where different economies are advised to specialize in those products that they can sell profitably in the international market.

The author describes a detailed study done on women's workers in the strawberry packing and freezing plants in Zamora, Mexico. The main reasons for employing women for this work are that they can be paid much lower wages than those stipulated by the law and that they accept conditions in which there is a constant fluctuation in schedules and days of work. Work in this agro-industry, for the majority of women, is certainly no way to get ahead, because there are no promotions, no education and training, and no prospects for improvement in the future. The "comparative advantage" for companies turns out to be a disadvantage for the women.

163. Ashby, Jacqueline A., and Stella Gomez. *Women, Agriculture, and Rural Development in Latin America*, Colombia: Centro Internacional de Agricultura Tropical, 1985.

Assesses the adequacy of the role of women in agriculture and food systems in Latin America for defining objectives and strategies in agricultural research programs. Key issues for agricultural technology research and development are outlined in terms of the effects of sex roles on food production and welfare of the rural poor in low income countries. Reviews the degree of participation of women in Latin American agriculture and what this implies for identifying women as special user groups for agricultural technology. The largest part of this book is the extended bibliography of 415 entries including studies in English and Spanish. Author and subject indexes are included.

164. Barlett, Peggy F. "Part-Time Farming: Saving the Farm or Saving the Life-Style." *Rural Sociology* 51:3 (1986): 289-313.

An in-depth study of a row-crop and livestock county in Georgia. The author finds that most part-time farmers are not "trying to save the family farm, nor are they hobby farmers." The majority of the farmers rejected full-time farming early in life and committed themselves to getting an education and a stable off-farm job. Most part-time farm wives in this study are much less involved in farm work than their husbands, and the reasons for this divergence are suggested in the distribution of life-style benefits and costs. An extended list of references is included.

165. Bokemeier, Janet L., et al. "Labor Force Participation of Metropolitan, Nonmetropolitan, and Farm Women: A Comparative Study." *Rural Sociology* 48:4 (1983): 515-539.

Looks at labor force participation of metro, non-farm-nonmetro, and farm women. The data gathered from a large statewide mail survey compare personal, socioeconomic, and family characteristics. The findings regarding correlates of labor force participation indicate that family and status are the most influential correlates of metro and nonmetro women's labor force participation, while status factors are more influential for farm women. References are included.

166. Boulding, Elise. "The Labor of U.S. Farm Women: A Knowledge Gap." *Sociology of Work and Occupations* 7:3 (August 1980): 261-290.

Points out that the census procedure may lead to serious undercounting of women's farm labor. In an exploratory study of 27 farm women from Oklahoma, Vermont, and Colorado, the author finds there is substantial involvement by women in farm work, but there is no clear relationship between involvement in the farm work and farm decision making.

167. Bourque, Susan C., and Kay Barbara Warren. *Women of the Andes: Patriarchy and Social Change in Two Peruvian Towns.* Ann Arbor, MI: The University of Michigan Press, 1981.

Describes the lives of women in two agricultural communities in the Andes: Mayobamba, a settlement of 450 people, and Chiuchin, a trade center with 250 residents on the mountainside north of Lima, Peru. The first part of the book examines women's position in the communities, particularly women's subordination, marriage, and family politics. The second part of the book discusses the sexual division of labor, the participation of women in a broad range of agricultural activities and the rigidly enforced cultural patterns governing access and sex role stereotypes. The authors also look at women's position in the agrarian class system, women's consciousness and participation in the patriarchal power structure, and the effect of social change and national development on the women in the two communities. A bibliography and an index are included.

168. Bronstein Audrey, *The Triple Struggle: Latin American Peasant Women.* Boston: South End Press, 1982.

From this account of conversations with women from different Latin American countries we get an idea of the magnitude of their problems. Issues of poverty, lack of resources, and injustice are relatively easy to recount, but more personal matters such as domestic tyrannies and blighted hopes, are more difficult to talk about. The book raises the question, do Third World women consider "bread and butter" issues to be their only concern?

The author lived several months in Guatemala, Peru, and Ecuador, and spent several weeks in El Salvador and Bolivia, in an effort to learn more about the issue of sex and class. She studied the oppression of peasant women in male-dominated societies in the Third World, the position of peasants and their victimization via underdevelopment as well as the unequal distribution of wealth within and between nations.

169. Brown, Judith K. "Economic Organization and the Position of Women among the Iroquois." *Ethnohistory* 17:3-4 (1970): 151-167.

The relationship between the position of women and their economic role is examined by comparing ethnohistoric and ethnographic data relating to the Iroquois of North America and the Bemba of Northern Rhodesia. It was found that the Iroquois women controlled the factors of agricultural production. The Iroquois agricultural activities, which yielded bountiful harvests, were highly organized under elected female leadership. The Iroquois women maintained the right to distribute and to dispense all food, even that produced by men. This was especially significant as stored food constituted one of the major forms of wealth for the tribe. This control of the Iroquois economy gave the women a high status in their society. References are included.

170. Bruner, Daina. "The Influence of the Women's Liberation Movement on the Lives of Canadian Farm Women." *Resources for Feminist Research* 14:3 (November 1985): 18-19.

Summarizes the events and effects the Women's Liberation Movement has had on rural women's lives in Canada. The Canadian Council on Rural Development (C.C.R.D.) implemented a survey in 1976 on the nature of work done by rural women, and the ten most important issues of concern to farm women were identified. The author points out that the Women's Liberation Movement has had two major influences on the lives of the Canadian farm women. First, the movement has generated a great deal of publicity on the status of women in general; second it has allowed Canadian farm women to see themselves in a new light. Studies indicate that farm women are now more vocal and more political.

171. Centro de Investigacion y Estudios de la Reforma Agraria (Nicaragua). *Tough Row to Hoe: Women in Nicaragua's Agricultural Cooperatives.* Rural Women's Research Team, Center for the Study of Agrarian Reform (CIERA), Managua, Nicaragua; San Francisco, CA: Food First, Institute for Food and Development Policy, 1985.

The Sandinista revolution in Nicaragua initiated an agrarian reform law in 1981, and it was the first time in Latin America that women were explicitly included as beneficiaries of reforms in the agricultural sector. This study analyzes the process by which women became members of the cooperatives and determines the extent of women's participation as a result of the agrarian reform law; what factors encourage women to participate in the cooperatives; and what were the barriers they encountered?

One of the more important findings was that as long as women carry the full burden of child care and housekeeping, they cannot participate in agriculture on an equal basis with men. The study illustrates the difficulties of integrating women into the cooperatives. One year after agrarian reform, only 6% of the cooperative members were women. However, the female members were able to enhance their technical, organizational, and political development.

172. Conte, Christine. "Ladies, Livestock, Land and Lucre: Women's Network and Social Status on the Western Navajo Reservation." *American Indian Quarterly* 6:1-2 (Spring/Summer 1982): 105-124.

Examines the role of Navajo women in a remote area of the western reservation. Looks at the process of production, consumption, and distribution of resources. The aim of the project is to compare women's economic strategies in two different socioeconomic settings: the reservation site and the integrated border town.

173. Deere, Carmen Diana. "Changing Social Relations of Production and Peruvian Peasant Women's Work." *Latin American Perspectives* 4:1-2 (Winter/Spring 1977): 48-69.

The central hypothesis of this paper is that the development of the productive forces under the transition from servile to capitalist relations of production is progressive in terms of the nature, duration, and intensity of women's work. The author explains that the development of capitalism, as an uneven dialectical process, has contradictory effects on different groups of rural women depending on their relation to the means of production. The case study material presented comes from the northern Peruvian Sierra department of Cajamarca.

174. Deere, Carmen Diana. "Cooperative Development and Women's Participation in the Nicaraguan Agrarian Reform." *American Journal of Agricultural Economics* 65 (December 1983): 1043-1048.

The Nicaraguan agrarian reform, which started with the Sandinista victory of July 1979, is unusual in two respects. First, the process of cooperative development has been the result of a large scale mobilization of peasant and rural workers by their own mass organization, and, second, the new cooperative members benefited through the agrarian reform, both men and women. This new agrarian reform law is the first in Latin America to establish the legal preconditions for the incorporation of a significant number of rural women. The cooperative census data on membership by sex are not yet available, but preliminary indications suggest that there are more women members than in any other Latin American reforms.

175. Deere, Carmen Diana, and Magdalena Leon de Leal. *Women in Andean Agriculture: Peasant Production and Rural Wage Employment in Colombia and Peru*. Geneva: International Labour Office, 1982.

The census data of the Andean region suggest that the participation of rural women in agriculture has decreased in recent decades, but the author implies that it may, in fact, have increased.
The study draws on a peasant household sample survey from three areas in Colombia and Peru. The census data indicate that in peasant smallholder agriculture, women participate in both field work and decisionmaking. Contains over 30 statistical tables.

176. Fassinger, Polly A., and Harry K. Schwarzweller. "The Work of Farm Women: A Midwest Study." *Research in Rural Sociology and Development* 1 (1984): 37-60.

This study was undertaken to explore how and to what extent farm women are involved in the organization and activities of contemporary family farms, and to assess variation in their work roles associated with size of farm operation. Suggests future direction for research on farm women's work. References are included.

177. Finlay, Barbara. *The Women of Azua: Work and Family in the Rural Dominican Republic.* New York: Praeger Publishers, 1989.

Based on field observation and surveys of women in six rural communities in the southwestern Dominican Republic, this study assesses the impact of women's employment in large agricultural companies, on various aspects of their lives, their status within their communities and families, and their hopes for the future. Two surveys were conducted: a representative sample of the women in the communities and a special sample of women employees at the large agribusiness food production facilities in the area. The major goal of the study was to provide answers to questions such as "What type of women are employed by the companies?" and "What is the impact on various aspects of the workers' lives, and families?" The questionnaires used are included in English and Spanish as well as the interview instructions, an index, and a list of references.

178. Garrett, Patricia M. *Some Structural Constraints on the Agricultural Activities of Women: The Chilean Hacienda.* Madison: Land Tenure Center, University of Wisconsin-Madison, 1976.

A conference on Women and Development held at Wellesley College in Massachusetts in June 1976, this paper analyzes some of the structural constraints on women as consequences of a particular interaction between a land tenure system and a form of family organization on the haciendas in Chile. Data are drawn from the post 1935 period as recorded in general and agricultural census. During this time women were disproportionately displaced from permanent resident employment on large estates, and they were not absorbed into temporary labor but they were increasingly confined to the smaller farms as unrenumerated family members. These trends are examined in some detail. Statistical tables are included.

179. Ghorayshi, Parvin. "The Indispensable Nature of Wives' Work for the Farm Family Enterprise." *Canadian Review of Sociology and Anthropology* 26:4 (1989): 571-593.

    This case study from St. Charles, Quebec, confirms the emphasis of other studies on the multi-dimensional role of the farm wife. The author calls for the identification of forces that have clouded our view of work in general and farm wives in particular, and indicates that any analysis of farming which omits or de-emphasizes wives' work presents a misleading view of agricultural farm structure. References are included.

180. Harry, Indra S., and Trevor A. Thorpe. "Agriculture in Trinidad: A Comparison of Male and Female Participants." *Social and Economic Studies* 39:2 (June 1990): 79-103.

    This study points out that men and women make equal labor contributions to agriculture in Trinidad. The data used represent part of a major exploratory study done in 1979. Information was obtained through a personally administered questionnaire survey, where 130 households were interviewed. Men were older and more educated than women, and many had off-farm employment. Land was often jointly owned, but the husband assumed responsibility for the farm and the crop selection. The data showed that, in general, men and women worked the same hours per day and per week. However, women worked longer hours in husbandry and men in sugar-cane production. The authors feel that the farm women in Trinidad should receive extension services, agricultural benefits and credits, appropriate training, and other national benefits which are normally given to men only.

181. Jones, Calvin, and Rachel A. Rosenfeld. *American Farm Women: Findings from a National Survey*. Chicago: National Opinion Research Center Report No.130, 1981.

    A nationwide survey of farm women was conducted by the National Opinion Research Center (NORC) with the following major objectives: to determine the extent of involvement in the work and management of their operation, membership in farm and community organizations, and participation in off-farm labor, farm women's experiences with the program activities of the U.S. Department of Agriculture and their perception of those experiences. The data were analyzed to identify those factors having greatest influence on women's involvement in their enterprise and in USDA programs. The farm women's responses were compared with those from a sample of male farm operators. Telephone interviews of 2,509 farm women and 569 men were conducted. The survey questionnaire, female and male versions, is included in the appendix.

182. Knudsen, Barbara, and Barbara A. Yates. *The Economic Role of Women in Small Scale Agriculture in the Eastern Caribbean: St. Lucia.* Women and Development Unit, Extra-Mural Development, University of West Indies, Barbados, Pinelands, St. Michael, Barbados: The Unit, 1981.

The purpose of this study is to ascertain the extent and nature of economic participation by women in small-scale agriculture in St. Lucia. The authors offer policies and strategies for the most efficient delivery of agricultural support services to improve the economic well-being of farm women and their families in St. Lucia. Research methods, questionnaires and references are included.

183. Lynch, Barbara Deutsch. "Women and Irrigation in Highland Peru." *Society and Natural Resources* 4:1 (1991): 37-52.

Despite norms that irrigation is men's work, the author found that women in many parts of the Andes now enjoy more control over water in the field than they did before. Most women who participate directly in irrigation and system maintenance projects are single women or wives of migrant workers. However, it is necessary for women to participate directly at the interface between the system and the agency in order to meet their needs as consumers of irrigated water.

184. Maret, Elizabeth and James H. Copp. "Some Recent Findings on the Economic Contributions of Farm Women." *The Rural Sociologist* 2:2 (1981): 112-115.

Reviews the conventional perception on economic activities of farm women and suggests future research is needed in several areas such as female-managed farms. Reviews the role farm women are playing in farm production and suggests that the degree of family farms survival in American agriculture may be due to the economic contribution of farm women working on and off the farm.

185. Odie-Ali, Stella. "Women in Agriculture: The Case of Guyana." *Social and Economic Studies* 35:2 (June 1986): 241-289.

Deals with women's role within the family and in agriculture, their knowledge of and access to credit and other agricultural development facilities which may be available, and their perception of themselves in the field of agriculture. Interviews with 62 women farmers and 10 men farmers took place and questions related to credit, land ownership, and problems were asked. Analysis of recent population censuses indicates a decline of female participation in agriculture, twice the extent of the decline in the male agricultural work force. This shows that Guyanese are moving away from agriculture at a time when Guyana should produce more food for survival. There is a need to arrest this declining trend in agricultural involvement and devise strategies of support networks in order to increase production levels and make farming as economically attractive as possible. References are included.

186. Ollenburger, Jane C., et al. "Labor Force Participation of Rural Farm, Rural Nonfarm, and Urban Women: A Panel Update." *Rural Sociology* 54:4 (1989): 533-550.

The authors trace the labor force participation of about 800 women in Nebraska from 1977 to 1985, examine changes in the work status of the cohort of Nebraska women during the farm crisis, then identify individual factors influencing labor force participation and continuity, contrasting three residential groups of women. They found that the farm women are now more likely than their nonfarm counterparts to participate in the paid labor force when they have high education levels and are single, divorced, or widowed.

187. Pescatello, Ann. "The Female in Ibero-America: An Essay on Research Bibliography and Research Directions." *Latin American Research Review* 7 (Summer 1972): 125-141.

Gives an overview of women's place in the Latin American society and the way Ibero-American scene is colored by its own peculiar mix of cultural antecedents and traditional perceptions. Also reviews new studies on women in Ibero-America; the bibliography is organized by country. Citations in English and Spanish are included.

188. Rosenfeld, Rachel Ann. *Farm Women: Work, Farm and Family in the United States*. Chapel Hill: The University of North Carolina Press, 1985.

Focuses on women who operate their own farms or are wives of farm operators. Looks at the nature of their farm and their families which shapes their work on and off the farm, with and without pay. The book is based on data from federal Census of Agriculture, and the author's research in the U.S. Department of Agriculture Farm Women's Survey in 1980. Telephone interviews were conducted with 2,509 female farm operators and 569 male farm operators to determine the extent of farm women's work, decision making, and off-farm employment. The data show that farm women's work changed according to the characteristics of the farm, the women's age and education, and the number of children in the family. Questionnaires used for the survey, and references are included.

189. Ross, Lois L. *Harvest of Opportunity: New Horizons for Farm Women*. Saskatoon, Saskatchewan: Western Producer Prairie Books, 1990.

Focuses on the work farm women are doing today, why they are doing it, and the obstacles they face while discovering and asserting their identities. Includes information on common concerns such as rural daycare, financial independence for farm women, business know-how, and the isolation of farm life. In all of these endeavors, the rural women have found a sense of self-worth and increasing self-confidence.

190. Rossini, Rosa Ester. "Women as Labor Force in Agriculture: The Case of the State of S. Paulo, Brazil." *Studi Emigrazione/Etude Migration* 20 (June 1983): 222-230.

When filling out census questionnaires, women omit references to the productive work they are performing, because they see their main occupation as "mother and housewife." This is how they become "invisible producers." The last census data in Brazil revealed a slight increase of female presence in the primary and secondary sectors of the economy and a marked increase in the tertiary sector.

191. Ruchwarger, Gary. *Struggling for Survival: Workers, Women and Class on a Nicaraguan Farm.* Boulder: Westview Press, 1989.

The author is the founder of a research program devoted to indepth case studies on Nicaraguan mass organization. Examines the economic context of a tobacco and vegetable state farm, including the class structure and the conflicts among workers, technicians, and managers. Also assesses the extent to which the feminization of the rural labor force and the organization of women workers have affected gender relations. Notes on recent theories on class and gender and a selected bibliography are included.

192. Russel, Scott C., and Mark B. McDonald. "The Economic Contributions of Women in a Rural Western Navajo Community." *American Indian Quarterly* 6:3-4 (Fall/Winter 1982): 262-282.

Data used to analyze the economic activities reported in this paper were gathered during research in Shonto, Arizona, in 1975 and 1976. Women have held, and continue to hold, an important economic role in animal husbandry activities, particularly sheep and goat husbandry. Detailed statistical information is included on income sources, herding patterns, arts and crafts activities, demographic and economic characteristics of weavers and potters, source of jobs held, and mean wage income. It was found that women's income level is much lower than men's.

193. Sachs, Carolyn E. "American Farm Women" in Stromberg, Ann H., et al., ed. *Women and Work: An Annual Review* 2 (1987): 233-248.

Women's activities in agricultural production in the United States have been overlooked. The sexual division of labor continuously changes with shifts in the economic structure of agriculture, and overall the major change over the last decades has been toward a decline in the number of both women and men involved in agriculture. Today, women work on both family farms and farms not owned by their families, and they participate in a wide diversity of farm tasks, household work, and off-farm employment. Women's labor on non-family farms is increasing relative to that of men. References are included.

194. Sachs, Carolyn E. *The Invisible Farmers: Women in Agricultural Production.* Totowa, NJ: Rowman & Allanheld, 1983.

Says "there is still a strong tendency to see men as farmers and women as farmers' wives" in the United States. As a consequence of a powerful sexual division of labor, women have often been consistently overlooked and undervalued. The primary purpose is to explore the full nature of women's involvement in agriculture production through the use of historical data and interviews with contemporary farm women. Points out that the male bias in the United States' system of agriculture has been transferred to developing countries through the many projects of the World Bank and USAID and that women's power and access to agricultural resources in the developing countries are lessened. Some recommendations and a bibliography are included.

195. Shaver, Frances M. "Women, Work and Transformation in Agricultural Production." *The Canadian Review of Sociology and Anthropology* 27:3 (August 1990): 341-356.

Examines the modernization of agriculture and the penetration of capitalist relations of production in Quebec, Canada, and then looks at how these changes affect women's participation in agriculture. Data were collected during a year of field work in a small parish in Quebec. Interviews were conducted with 63 randomly selected farm families. The findings indicate that neither the development of capitalist agriculture nor the adoption of modern methods has an independent effect on women's contribution. However, the study indicates that modernization will only displace women when non-capitalist relations of production are replaced with capitalist ones and then only with respect to farm work.

196. Shortall, Sally. "Canadian and Irish Farm Women: Some Similarities, Differences and Comments." *The Canadian Review of Sociology and Anthropology* 30:2 (May 1993): 172-190.

Considers the similarities between farm work, farm household work, child care work, and voluntary community work carried out by Canadian and Irish farm women and how in both cases it is undervalued. The major differences between the two sets of farm women is the organization of Canadian farm women into farm women's groups, most of whom are affiliated to the Canadian Farm Women's Network. These groups have successfully increased the attention given to the role of farm women and their work. There is no comparable group of farm women in Ireland, and the reason for this is explored. The contribution of farm women's groups to comprehensive rural development policy is also considered. List of references is included.

197. Straus, Murray A. "The Role of the Wife in the Settlement of the Columbia Basin Project." *Marriage and Family Living* 20 (February 1958): 59-64.

Examines the relationship of selected characteristics of Columbia Basin Project settlers' wives to success in developing their farm into a profitable and personally satisfying venture. Compares the wives of a low and a high success sample of families and looks at background characteristics and tangible contributions as well as intangible contributions. Concludes that the data available fail to support expectations concerning the relation of wives' direct economic contribution to success in new settlement. Instead, the qualities which differentiated the high- from the low-success group are certain attitudes, values, and personal characteristics.

198. Straus, Murray A. "Family Role Differentiation and Technological Change in Farming." *Rural Sociology* 25 (June 1960): 219-228.

To test the hypothesis that farm operator technological competence is associated with an "integrative-supportive" wife marital role, the author used a sample of 903 Wisconsin farm operators who were classified into high and low technological competence groups. Using the farm practice adoption index, the highest scoring 27% were selected as the operators representing high levels of technological competence and the lowest scoring 27% as those representing low levels of technological competence. He then looked at the background characteristics of the two groups of wives, such as age, education, home management, and decision making. Altogether 46 items were used to test the hypothesis, and 15 of these items were found to differentiate the two groups of wives significantly.

199. Sweet, James A. "The Employment of Rural Farm Wives." *Rural Sociology* 37:4 (1972): 553-577.

The employment patterns of rural farm wives are examined by using the 1960 census data. Multivariate analysis of both the probability of being currently employed in a nonfarm job and the probability of having received salary income during 1959 is presented, and the employment differentials among rural farm wives are compared with those among urban wives. An attempt is made to differentiate the farm population according to husband's occupation, tenure of housing, and source of income and to examine differential patterns of wife's employment. The author found that rural farm wives are employed in large and growing proportions and that the variation in employment rates of rural farm women is, in general, quite similar to the variation among urban women.

200. Valentine, Frances Wadsworth. *Successful Practices in the Employment of Nonfarm Women on Farms in the Northeastern States, 1943*. Bulletin of the Women's Bureau, No. 199. Washington: United States Government Printing Office, 1944.

Describes the agriculture of the Northeastern region of the United States and discusses the labor demand and the farms on which the women worked, how the women were recruited, the farm work and working conditions in 1943 as well as life and living conditions of these women. It concludes with advisable practices for 1944.

201. Wilson, Fiona. "Women and Agricultural Change in Latin America: Some Concepts Guiding Research." *World Development* 13 (September 1985): 1017-1035.

It is not possible to draw simple conclusions from the existing research literature of how the lives of rural women are affected by the process of commercialization and capitalization in agriculture. "Researchers working on this theme must take a position on what they understand by agrarian change along capitalist lines; how they conceptualize women's position in society and relation of gender; and how they relate changes in agriculture and the lives of rural women," says Wilson. The author looks at the underlying conceptual and political differences among commentators, discusses the basic weakness of the "impact model" most commonly adopted up to now, and indicates directions taken by alternative conceptualizations that stress the interplay of relations of gender with relations of class and ethnicity. Discusses also the impact of capital's intervention in agriculture on women, the process of capitalist change and rural women's productive and reproductive roles as well as the process of capitalist change and women's position within relations of inequality.

202. Young, Kate. "Changing Economic Roles of Women in Two Rural Mexican Communities." *Sociologia Ruralis* 18:2-3 (1978): 197-214.

Looks at the effects of incorporation of two small communities into the national economy and the changes this had on women's positions. In both communities the women were weavers providing cloth for domestic needs and for the market.
With the collapse of cotton cloth production and the introduction of coffee cultivation into one of the two villages, their history diverged. Outlines the changes as well as the differences and argues that although the roles for women have diversified to a greater degree in one community than the other, the effect on certain categories of women in both communities is similar.

ASIA AND AUSTRALIA

203. Agarwal, Bina. "Impact of Rural Development on Economic Status of Women." *Indian Journal of Agricultural Economics* 40:3 (1985): 282-290.

This report examines four interrelated aspects that require particular focus: (1) assessment of women's economic status relative to men's on the basis of specific criteria, (2) changes of women's economic status as a result of agricultural and rural modernization, (3) impact of government's anti-poverty and related programs on women's economic status, (4) wider implications of women's economic status, especially for the physical well-being and social status of women and female children.

There were 33 papers accepted for discussion and 19 of these focused on some issues concerning the first aspect; 9 dealt with specific part of the second aspect; 4 with the third; and only 1 with the fourth aspect. The rapporteur was disappointed with the papers because the majority were descriptive and not analytical, and most have some methodological weakness. Makes critical comments.

204. Aimed, Mayan Recognition. "Unseen Workers: A Sociocultural Profile of Women in Bangladesh Agriculture." *Society and Natural Resources* 5:4 (1992)): 375-390.

The male-dominated Bangladesh society undervalues women's work by paying them less than men with the same workload and the same output. Credit is often taken by the women's husbands. This essay explores the possibilities of policies that could recognize women's labor and incorporate women in comprehensive development activities throughout Bangladesh.

205. Atal, Yogesh, ed. *Women in the Villages, Men in the Towns.* Paris: UNESCO, 1984.

National and international agencies have begun paying more attention to the problems of women and have initiated several programs of actions to facilitate their equal participation in the development process. This book presents a summary of the programs examining problems faced by women in Asia. Each country study covers the following topics: (1) country profile and description of the changing situation, (2) characteristics of the rural migrant, (3) the women in the family left behind (this is the prime focus of the studies which analyze the variety of roles performed by women when men are absent), (4) intensive case studies of the villages involved, (5) community adjustment, (6) case studies on family structure and communal lives of women. The countries covered are: Bangladesh, India, Philippines, and Korea.

206. Croll, Elisabeth J. *Women in Rural Development: The People's Republic of China.* Geneva: International Labour Office, 1979.

Looks at women's participation in the subsistence sector and family agricultural production, women's participation in wage labor, and the impact of agricultural organizations on rural women as well as women's reproductive activity, and its bearing on women's participation in productive work. Also analyses the impact of various influences and policy measures on the situation of rural women in China. Concludes with a summary of the main policy areas requiring attention and the relevance of the Chinese experience for other countries and highlights the point that even a revolutionary transformation of socio-economic structures may not suffice to bring about a state of equality for women. Some statistical tables are included.

207. Evenson, Robert. "Time Allocation in Rural Philippine Households." *American Journal of Agricultural Economics* 60 (1978): 322-330.

Reports on an analysis of time allocation data collected from rural Philippine households as part of a larger research effort on agrarian household behavior. Develops a simple analysis of factors determining time allocation of household members and reports a simple econometrics analysis of the data from the Philippines.

208. Hart, Gillian. "Productivity, Poverty, and Population Pressure: Female Labor Development in Rice Production in Java and Bangladesh." *American Journal of Agricultural Economics* 65:5 (December 1983): 1037-1042.

The agricultural production of rice in Java and Bangladesh has been very different. The Javanese rice sector has been a great success. The main sources of growth have been fertilizer and irrigation expansion rather than the intensification of labor use per hectare. This study seeks to show how the mobilization and deployment of female labor is crucial to understand the differences in the performance of the rice sector in the two countries and to assess the distributional implications of changing patterns of rice production.

209. Heyzer, Noeleen, ed. *Women Farmers and Rural Change in Asia: Towards Equal Access and Participation*. Kuala Lumpur: Asian and Pacific Development Center, 1987.

Provides examples of the range and diversity of rural development in several Asian countries and the impact it has on women. Assesses the framework within which development plans and policies for rural women have been formulated in China and India. Another chapter examines the effects of colonialism and post-independence rural programs on rural areas in Malaysia. We also find chapters focusing specifically on large-scale rural development projects such as hydroelectric dams, irrigation, and land settlement schemes. The final chapter looks at the impact of China's rural reforms from the communal to the household contract system on the work patterns and status of women. Bibliographies after each chapter are included.

210. Hussain, Sahadad, M.D., et al., *Women's Contribution to Homestead Agricultural Production Systems in Bangladesh*. Comilla, Bangladesh: Bangladesh Academy for Rural Development, 1988.

A sample of one hundred farm families from five different farm categories, landless, marginal, small, medium, and large from each location was selected by stratified random sampling technique. The following categories were studied: socio-economic profile; ownership pattern of homestead and cultivated land; status of forest, fruit and vegetable production in homesteads; crop and vegetable seed management; post harvest processing; livestock production and utilization systems; backyard pond utilization; food production, dietary habits and nutrition; agricultural information and training needs; participation of women in decision making process; constraints in homestead agriculture. Conclusions and recommendations are included. Statistical tables are included after each chapter with a short list of references, and a list of scientific names of different plant species and a glossary are also included.

211. Ireson, Carol. "Women's Forest Work in Laos." *Society and Natural Resources* 4:1 (1991): 23-36.

In this study 120 women farmer/gatherers were interviewed from eight villages in Bolikhamsai Province in Central Laos. Observation, forest visits, meetings, and conversations provided additional information. It was found that women with access to old-growth forest as well as second-growth areas use forest products mainly for subsistence purposes whereas women with access only to second growth areas are more commercially oriented and are more likely to sell what they gathered. It is clear that women's forest work contributes regularly to the economy of their households.

212. Jacobson, Doranne. "Indian Women in Processes of Development." *Journal of International Affairs* 30:2 (1976-77): 211-242.

A general overview of women's role in the ongoing process of Indian national development is presented. Women agricultural laborers are 92% illiterate and suffer from low pay and wage discrimination. Men monopolize the new agricultural methods and machinery, and women continue to use the more laborious tradition methods or suffer unemployment.

213. Johnson, Marshall, et al., "Chinese Women, Rural Society and External Markets." *Economic Development and Cultural Change* 35:2 (1986): 257-278.

Attempts to incorporate China's experience during the first half of this century into larger discussions of how market exposure occasioned by foreign contact affects the position of women and other social groups in developing societies. The data for this exercise involve a reanalysis of a farm survey collected by students and faculty at Nanking (Jingling) University around 1930. It appears that market exposure tends to increase rural labor demand, raise rural wages, and leave social inequality unchanged. Compared to conditions in China today women were left behind in agriculture while their men left to work in the cities.

214. Johnson, Patricia Lyons. "Women and Development: A Highland New Guinea Example." *Human Ecology* 16:2 (1988): 105-122.

Report of a study on household variables and their relationship in cash cropping. The data were gathered among the Gainj of Madang Province, Papua New Guinea. Comparison of data from Census and household surveys from 1978, the year in which cash cropping began, and a survey of 110 households from ten Gainj parishes in 1983. The results illustrate the importance of women's labor in economic development and the dynamic nature of the relationship between household structure and economic development.

215. Judd, Ellen. "Alternative Development Strategies for Women in Rural China." *Development and Change* 2:1 (January 1990): 23-42.

   The data for this study were collected in three villages in Shandong during the summer of 1986 and winter 1987. The author included interviews and surveys. Three major strategies for enhancing women's role in economic development in rural China are in evidence in the 1980s: (1) replacing male labor in agriculture, (2) employment of women in rural industry, (3) commodity production, where the women independently produce goods and services for the market.

216. Khan, Zarina Rahman. "Women's Economic Role: Insights from a Village in Bangladesh." *Journal of Social Studies* 30 (1985): 13-25.

   In Bangladesh there is a strict division of labor along gender lines. The areas of activity for men are the productive and public ones, while those of women are domestic and private. Only men's work is valued in cash terms or economically and they are considered superior, and have responsibility for the women. The author analyses the traditional and non-traditional work of women in a study conducted in a village in the Dhaka District, where 167 households participated.

217. Lyson, Thomas A. "A Note on the Increase of Female Farmers in the United States and New Zealand." *Australian and New Zealand Journal of Sociology* 26:1 (March 1990): 56-67.

   Looks at the literature on the roles of women in agricultural production in modern societies. The author says it has taken on the aura of detective stories; since the contribution women make to farming is hidden, invisible, neglected, or underestimated, the task of the researcher has been to uncover the range of tasks women perform on the farm. In the United States the number of female farmers increased across all age groups between 1970 and 1980, and similar results were obtained in New Zealand. The largest increase in both countries was among a relatively young age group.
   It must be noted, however, that agricultural trade is the life blood of the New Zealand economy, and the agricultural industry is growing at a rate

that is four and one half times greater than in the United States. The number of production agricultural workers increased 12.9% between 1976 and 1981 in New Zealand, while the United States registered a 9% decrease in farmers, farm managers, and farm workers between 1970 and 1980.

218. MacPhail, Fiona, and Paul Bowles. "Technical Change and Intra-Household Welfare: A Case Study of Irrigated Rice Production in South Sulawesi, Indonesia." *The Journal of Development Studies* 26 (October 1989): 58-80.

The differences in male and female employment levels under rain-fed and irrigated rice production are investigated in a case study of two villages in South Sulawesi, Indonesia. Existing theories of the household suggest that the distribution of resources within the household is a function of household members' employment. It is concluded that the pattern of labor allocation in the irrigated area involves higher absolute labor requirements for both men and women, compared to the rain-fed area, but that women's agricultural labor relative to men's is lower. It is argued that the differences can be seen as a response to the different agricultural production technologies used, and it suggests that the levels of inequality within the household are greater in the irrigated than in the rainfed area. Several statistical tables and a list of references are included.

219. Mencher, Joan P., et al. "Women in Rice Cultivation: Some Research Tools." *Studies in Family Planning* 10:11-12 (November/December 1979): 408-412.

Documents the role that women in India play in rice cultivation. The women are employed as agricultural laborers, supervisors in family-owned fields, employers of labor, and participants in the agricultural decision-making process. The most innovative part of the study is the charts the authors created and used with the illiterate women who work in the fields.

220. Mencher, Joan P. and K. Saradamoni. "Muddy Feet, Dirty Hands: Rice Production and Female Agricultural Labour." *Economic and Political Weekly* 17:52 (December 1982): 149-167.

Presents a detailed study on the involvement of women in rice production in six villages in the states of Kerala, Tamil Nadu, and West Bengal in India. The authors give a historical overview of the rice cultivation and look at female-headed households, discrimination in wages, displacement of female agricultural laborers, and alternative employment. A detailed description of the methods used in collecting the data and an analysis of the data are included. Statistical tables and a short list of references are included.

221. Miller, Barbara D. "Female Labor Participation and Female Seclusion in Rural India: A Regional Review." *Economic Development and Cultural Change* 30:4 (1982): 777-794.

Looks at the variation in both form and formality of female seclusion and segregation in the North and South of India. The author found that female seclusion and segregation can be found throughout India, but in the North the seclusion of females is more pervasive and practiced more strictly. In the South there is much formal avoidance between certain males and females at specific points in the life cycle, but the actual confinement of females within the home is restricted to a few elite groups. Census of India data are used to construct a rough picture of the regional pattern of female labor participation.

222. Nelson, Nici. *Why Has Development Neglected Rural Women? A Review of the South Asian Literature.* Oxford, New York: Pergamon Press, 1979.

This book is part of the Women in Development Series. It deals with global economic development strategies and female emancipation. The author presents a review of the literature and an overview of the studies on the role of women in rural development in four countries in South Asia: Bangladesh, India, Pakistan, and Sri Lanka. It includes an extensive bibliography.

223. Panter-Brick, Catherine. "Motherhood and Subsistence Work: The Tamang of Rural Nepal." *Human Ecology* 17:2 (1989): 205-228.

Examines the work patterns of Nepalese rural women of different childbearing status and the degree to which women modified their role in the local economy during pregnancy and lactation. Fieldwork was undertaken in a remote village in the foothills of the Himalaya in the village of Salme. Time allocation and energy expenditure were studied over a period of one year in 1982-1983. Observations were completed by a food survey, regular anthropometric measurements, collection of blood and stool samples, and records of household socio-economic change. A total of 43 Tamang women (and their male kin) were involved in the study.

It shows that combining work with child care is conditional upon a particular combination of ecological and socio-demographic characteristics such as the seasonal demand for labor, the geographical dispersion of family members in nuclear families, long birth intervals, and an egalitarian and flexible distribution of work. Any distortions of these conditions would alter the balance and decisions to make the best use of resources. References are included.

224. Sajogyo, Pudjiwati et al., "Studying Rural Women in West Java." *Studies in Family Planning* 10:11-12 (1979): 364-370.

Describes the problems of rural women in the household, in the labor market, and in society. The author discusses how to achieve a better understanding of the causes of these problems and how to identify policies and programs at the national, regional, and local level. Describes also the type of information sought and the research methods used as well as analysis and presentation.

225. Saradamoni, K. "Women's Status in Changing Agrarian Relations: A Kerala Experience." *Economic and Political Weekly* 17:5 (January 30, 1982): 155-162.

Changes in agrarian structure in India brought about changes in the lives and status of women. Examines the progressive breakdown of landlordism in the Palghat district in Kerala and the socio-cultural changes that took place that affected women belonging to different strata of the agricultural community. Points out there is a need to look at development as a process of change and to understand the contradictions that emerge during the process. The progressive advancement in land ownership and tenant system turned out to be a gradual slide-back for the women.

226. Sato, Kunio. "Wives of Farmers, a Critical Import." *Japan Quarterly* 35:3 (1988): 253-259.

Japan's farming villages suffer a critical shortage of brides. This is a story of six women from Santiago, north of Manilla, who were chosen out of 400 respondents, to come to Higashiiyayama on the island of Shikoku in Japan and who married and settled down as farmers' wives.

227. Schoustra-van Beukering, E.J.E. "Sketch of the Daily Life of a Bengali Village Woman." *Plural Societies* 6:4 (1975): 51-66.

Focuses upon the daily life of women, their place in the household, in the kinship group, and in the wider social context. Agriculture is of basic importance: it is dominated by paddy (subsistence economy) and the cash crop jute. Gives detailed information on the relationship of women in the patrilineal extended family and the working schedule of the village housewife.

228. Scott, Gloria L., and Marilyn Carr. *The Impact of Technology Choice on Rural Women in Bangladesh: Problems and Opportunities.* World Bank Staff Working Papers No. 731, Washington, D.C.: The World Bank, 1985.

Provides background on the social, cultural, and economic issues relevant to the concern for increasing women's opportunities for employment and income generation in rural Bangladesh. Describes the full range of women's economic contribution and importance to the rural economy and to their families. Suggests approaches for increasing income-earning opportunities and improving the status and skills of women. References are included.

229. Sharma, Ursula M. "Women's Participation in Agriculture in India." *Current Anthropology* 23:2 (April 1982): 194-195.

Explores the nature of female roles in rural Indian society through an investigation of the part women play in production. Fieldwork was conducted in two villages in 1977-78 in Punjab, where development in agriculture has been very rapid and women today play a relatively small role in agriculture production; and in the lower foothills of Himachal Pradesh, where agricultural development has been slow and women play a greater role than before because of the absence from home of large numbers of male migrants.

230. Singh, Andrea M., and Nerra Burra, eds. *Women and Wasteland Development in India*. New Delhi: Sage Publications, 1993.

Papers presented at an ILO National Technical Workshop on Women and Wasteland Development in 1991. Martha Chen's background paper presents a rationale for women's participation in wasteland development projects in India and traces the origins and development of government policies regarding this development. Includes 20 projects in eight different states covering several different agro-ecological regions. Several case studies are enclosed, and most of the authors argue that women's needs and priorities are different from those of men and wasteland development should be viewed as part of an integrated human and natural resource development strategy with the objective of creating a better overall quality of life for women, men, and children. A list of the contributors and an index are included.

231. Smith-Sreen, Poonam, and John Smith-Sreen. "Insight from Women Dairy Farmers in India: What Do They Gain from Participation?" *Social Action (New Delhi)* 41:4 (1991): 416-427.

Highlights all the benefits occurring from dairy farming as perceived by women dairy farmers who were interviewed in three regions of India. In dairying the women are involved in all the operations from feeding to the preparation of milk products and marketing, as well as acting as active members of the cooperatives. The authors are specifically interested in what impact this participation had on the lives of the women and whether it has been a positive change. The benefits as perceived by the women farmers are supplementary household income, dung fertilizer for household farms, improved nutrition for the family; source of regular income; improved prestige in the community, increased self-reliance and self-confidence, and decreased dependence on local moneylenders. On the other side these chores took from 6 to 8 hours of work each day, and the women had no say in how the money was spent although they were responsible for earning it.

232. Stoeckel, John, and N. L. Sirisena. "Gender-Specific Socioeconomic Impacts of Development Programs in Sri Lanka." *The Journal of Developing Areas* 23:1 (1988): 31-42.

An analysis of the comparative impacts of three national development programs and their combinations upon the occupational and income status of females and males in Sri Lanka. Utilizes data from a notional sample survey of the country, including 3,597 households stratified on the basis of development program areas. The authors found that the magnitude of total development program effects upon income in Sri Lanka is gender specific. All programs make a greater contribution to increasing male income than they do to increasing female income. The study showed that it is not sufficient for development programs to create female employment outside of the home to affect income; it is the type of employment that is crucial for income generation.

233. Stoler, Ann. "Class Structure and Female Autonomy in Rural Java." *Signs: Journal of Women in Culture and Society* 3:11 (1977): 74-89.

The author, who carried out field work in a Javanese village, considers two questions in this paper: first, by what means and to what extent have women gained and maintained access to economic independence and social power in Javanese rural society, and second, what are the effects of increasing demographic pressure and economic stratification on production relations and the role women play within them? Gives a historical perspective of the sexual division of labor in Java and examines the peasant class structure and the economic role of women, rice harvesting, the Javanese market system, and domestic production and interhousehold exchange. References are included.

234. Susheela, H., et al., "Women in Agriculture under Different Landholdings." *Indian Journal of Social Research* 32:3 (1991): 272-276.

The present study was undertaken by AICRP Home and Farm Management in four villages of Dharwad Taluk in Karnataka State in India. A total of 206 households were selected by stratified random sampling technique under different land holdings categories. It was found that more than half the population of women participated in agriculture in various capacities. Further studies are needed to learn more about women with regard to their occupation, the problems they face and the need for their development.

235. Consultative Workshop on Women in Rice Farming Systems in the Philippines (1987: University of the Philippines at Los Banos). *Filipino Women in Rice Farming Systems.* Los Banos, Philippines: International Rice Research Institute, 1988.

Presents selected papers discussed in the Consultative Workshop on Women in Rice Farming Systems in the Philippines held at the continuing Education Center of the University of the Philippines at Los Banos in March 1987. Aims to develop mechanisms by which women's role and needs will be considered at appropriate stages of technology development and dissemination. Also examines the effectiveness of the agricultural research and extension process. Specific project development activities began for this program in August 1985.
The chapters discuss projects on women in rice farming, arrowroot production and processing, women and root crop technology, participation of rural women and children in hand-watering agriculture, integrated pest management, training the village non-traditional extension audiences, and the role of women in co-operation. References after each chapter, many tables, and a list of authors and participants are included.

236. Waghmare, S.K., and Chaudhari, N.V. *Tribal Women in Agriculture.* New Delhi: Metropolitan Book Co., 1989.

Limited to ten villages in the Dangs district of Gujarat State and to activities of tribal women in relation to overall participation in crop husbandry, animal husbandry, poultry keeping, and home activities, the study explores the personal and socio-economic characteristics of the tribal women, the training needs, and the constraints faced by these women. From each sample village, 20 respondents were selected randomly to make the sample size of 200 respondents. Reviews of the literature and the research methodology are provided, and many statistical tables are included in the chapter on "Findings and Discussion."

237. White, Christine Pelzer. "Collectives and the Status of Women: The Vietnamese Experience." *Convergence: An International Journal of Adult Education* 17:1 (1984):46-54.

The Vietnamese peasant women became active members of the agricultural cooperatives on an equal footing with men, says a well-documented study from Hanoi.

Research carried out in 33 villages in Vinh Phu Province shows that women not only performed those steps of rice production termed "women's work," such as cultivation and weeding, transplanting seedlings, and harvesting, but they also did plowing and harrowing, which are considered "men's work." In addition, they also worked in tea cultivation and maize production. In the initial stages of forming the cooperatives, men tended to be more reluctant to join and quicker to leave than their wives because of the threat to their independent status as the household head of a family farm. Women gained help with the particular problems of pregnancy and childbirth as well as the care of young children, and the redistribution of labor made it possible for them to participate in education and training.

There is no question that rural collective institutions have improved women's status, but there still are some problems. For example, the author says men tended to receive more points for a day's work than women, and Vietnamese men have continued to have more time than their wives to devote to education, training, and political activities.

238. Wilson-Moore, Margot. "Women's Work in Homestead Gardens: Subsistence, Patriarchy, and Status in Northwest Bangladesh." *Urban Anthropology* 18 (Fall/Winter 1989): 281-297.

The study, based on fieldwork in Chuchuli, a village in Northwestern Bangladesh, illustrates how the success of women in managing homestead gardens may not positively affect their socio-economic status. The income generated by the sale of garden produce is handled by men of the household rather than by the women, specially in the circumstances in which both Muslim and Hindu women are restricted in their participation in the local marketplace.

239. Xiyi, Huang. "Changes in the Economic Status of Rural Women in Transformation of Modern Chinese Society." *Social Sciences in China* 13:1 (January 1992): 83-105.

Following the economic revival and completion of land reform from 1950 to 1952 the Communist Party of China and the Central People's Government put forward a plan to bring about gradual socialist industrialization and achieve a step by step socialist transformation of agriculture.

As far as rural women are concerned, the deepest impact on their lives has been adjustment and consolidation of their economic status. Women's sphere of participation broadened to cover almost all projects and activities previously handled by men. By 1988, household specialization in agriculture, forestry, and animal husbandry led by women accounted for approximately 40% of households. The author also looks at sexual division of labor and presents data from a survey made in the rural areas of Miyi County, Sichuan Province in 1988, and she points out how rural women's status has been established and regulated, which is an event of historical significance both for the Chinese society and for the women themselves. Compared with men, the transformation of their economic status has been difficult and slow, and the women have paid a much higher price for it. This article was translated by Lyn Jeffery from *Shehuixue Yanjiu*, 1990, No. 6.

240. Youssef, Nadia H. "Women and Agricultural Production in Muslim Societies." *Studies in Comparative International Development* 12:1 (Spring 1977): 41-58.

Examines the scant information that exists about women in the agricultural development in Muslim countries: the specific types of agricultural tasks assigned to women, whether or not specific types of programs are needed to provide for, expand, or as the case may be, reduce the employment of women in agricultural labor; and whether or not specific types of training are needed to provide the proper skills for the various development programs of the future. States that if the women can be placed outside of the family production system, a contradiction between women's economic role and their family role (reproduction) can emerge. In this way,
the employability of women in rural areas could achieve three major purposes related to development strategy: (1) enhance economic productivity in rural areas in nonagricultural spheres, (2) decrease the fertility rate in rural settings, (3) strengthen the status and role of women as they increasingly come to identify and are, in turn, identified by others in terms of activities that they can legitimately achieve as independent producers and income

earners outside of their traditional roles of wife and mother. Several statistical tables are included.

EUROPE

241. Barbic, Ana. "Farm Women in Slovenia: Endeavors for Equality." *Agriculture and Human Values* 10:4 (Fall 1993): 13-25.

Gives an overview of farm women's issues in western and former socialist countries and identifies more similarities than differences between the regions. Detailed description of farm women in Slovenia based on extensive research serves as a database on how women's situation within the family, the farm, and in public life can be improved, and how gender equality as a final goal of rural women can be achieved. Statistical tables and a few references are included.

242. Barbic, Ana. "Women's Issues in Rural Europe" in Daniel Thorniley, ed., *The Economics and Sociology of Rural Communities: East-West Pespectives*. Brookfield, VT: Avebury, 1987, pp.124-162.

Presented at a seminar in Balatonfoldvar, Hungary, in September 1985, the author gives the facts about rural women's situation in Europe as agricultural producers, the legal and economic position of women farmers, social security, and similarities and differences between western and socialist countries. She also looks at strategies for emancipation of rural women in this area and what the future will be. References are included.

243. Berlan, Martine. "Farmers' Wives in Protest; A Theater of Contradictions." *Sociologia Ruralis* 26:3-4 (1986): 285-303.

Explains public protest actions, a strategy adopted by farm women in France in their efforts to enter the political sphere, and examines the farm women's role in challenging the social categories of gender. From the research that has been done, it seems that these actions express a strategy of resistance and defense of the family farm rather than an autonomous feminine social movement in agriculture.

244. Besteman, Catherine. "Economic Strategies of Farming Households in Penabranca, Portugal." *Economic Development and Cultural Change* 38:1 (October 1989): 129-143.

    Conducted fieldwork on household strategies of farming households in 1984 in a rural parish in northwest Portugal. The farming community consisted of two kinds of farms those where the husband and wife were both full time farmers, referred to as joint farms, and those where the wife was the full time farmer and the husband was employed off-farm either locally or abroad--referred to as female operated farms. The study demonstrates how increased off-farm employment opportunities for men affected the household strategies of farm households. The female-operated farms represented over one-third of the farms in the parish, and the author points out how the extra income has permitted some households to acquire farms, and others to modernize their farms.

245. Bridger, Susan. *Women in the Soviet Countryside: Women's Role in Rural Development in the Soviet Union.* Cambridge: Cambridge University Press, 1987.

    Gives a history of rural women in the development of the Soviet Union. The old peasant family has vanished, and in the rural family of 1987, women tend to be better educated than their husbands and often take the lead in decision making. The process of modernization is one over which rural women themselves had little control. Development planning remained in the hands of the Communist party, in which rural women were always poorly represented. The book is divided into three parts with the first part discussing women in the rural work force, the effect of the rural population's migration to the cities, women as machine operators and dairy farmers, and women in management. The second part looks at women in the rural family, emergence of smaller families, division of labor, parental influence in marriage, and inequality within the marriage. In the final part of the book, the author includes information on women's role in rural culture; education of rural women, women's religious beliefs; and female participation in political activities. Each chapter has appropriate statistical tables, a bibliography, and an index are included.

246. Cernea, Michael. "Macrosocial Change, Feminization of Agriculture and Peasant Women's Threefold Economic Role." *Sociologia Ruralis* 18:2-3 (1978): 107-122.

Points out that feminization of agriculture in Romania is impressive. The structural changes in agriculture in the last 25 years with overall collectivization of agriculture, rapid broadening of industrial employment opportunities, and heavy rural-urban migration, all directly affected peasant women. The sex ratio of the labor force employed by Agricultural Producer Cooperatives in 1974 was 36.8% men and 63.2% women. Discusses and challenges the widespread thought that agriculture is typically a male occupation.

247. Fedorova, M. "The Utilization of Female Labor in Agriculture" in Gayle W. Lapidus, ed., *Women, Work and Family in the Soviet Union.* Armonk, N.J.: M.E. Sharpe, Inc., 1982, pp.131-146.

Analysis of general trends in the development of production indicates that the proportion of women working in agriculture is on the average higher than the proportion of women working in the national economy. However, the training and employment of skilled workers, especially the vocational training of women in the countryside, do not yet meet the requirements of scientific and technological progress.

248. First-Dilic, Ruza. "The Productive Roles of Farm Women in Yugoslavia." *Sociologia Ruralis* 18:2-3 (1978): 125-138.

Examines the productive roles of farm women in Yugoslavia and evaluates the working relations between the sexes in the rural development of Yugoslavia. The analysis of agricultural census data reveals that in socialist farming, women substituted for men who left the farm, and their only opportunity for competition was in the cooperatives, where they participated in self-management projects, and that did not happen very often. The labor force on part-time family farms was more feminized than full-time farms. Decision making on the farm was only influenced by the women when family life or household investments were concerned. Farm production and farm sale decisions were only made by men.

249. Garcia-Ramon, Maria Dolors and Gemma Canoves. "The Role of Women on the Family Farm: The Case of Catalonia." *Sociologia Ruralis* 28:4 (1988): 263-270.

>Describes the result of research carried out in a Catalonian village on the nature of women's work. The authors used agrarian census data, which they say are a poor source of information for analyzing women's contribution to agricultural activities. However, some conclusions were drawn. Nine in-depth interviews were also conducted and offer a closer and more reliable picture of women's contribution to agriculture.

250. Gasson, Ruth, et al. "The Farm as a Family Business: A Review." *Journal of Agricultural Economics* 39:1 (1988): 1-41.

>A great majority of farms in the United Kingdom are run as a family business, but the dimensions of these businesses are often neglected. Reviews the insights from social anthropology, history, and rural sociology as well as industrial economics as they apply to the farm family business. Relates these insights to considerations of the family development cycle, processes of inheritance and succession, roles of farmers' wives, and multiple-job farming families. Family relationships may have become less relevant in the case of smaller farms, but more relevant to the conduct of a successful large farm business. An extended bibliography is included.

251. Gasson, Ruth. "Farm Wives: Their Contribution to the Farm Business." *Agricultural Economics* 43:1 (1992): 74-87.

>Describes the nature of the work that the wives of the farmers do in the United Kingdom and Ireland. The author uses data obtained through two recent postal surveys as well as data from the Farm Business Survey for England and Wales. The largest group of women involved in farming are women married to farmers, and the author found that the importance of the wives in the labor force is increasing.

252. Haugen, Marit S. "Female Farmers in Norwegian Agriculture: From Traditional Farm Women to Professional Farmers." *Sociologia Ruralis* 30:2 (1990): 197-209.

A new amendment to the act regulating the right of succession to a farm upon the death of the owner passed in 1974. The new amendment gives the right to take over the farm intact to the eldest child, irrespective of gender.

Data for this study is based on interviews with 90 female farmers in 9 communities, and a control sample of 72 male farmers from the same communities were also interviewed. To assess if differences between the youngest and the oldest groups of female farmers were due primarily to a change in female attitude and practice or to a general generational change, the female sample was compared with the male sample. It was found that the female farmers represent two different categories, a traditional and a modern, depending upon the age of the farmer. Female farmers below 40 years of age were much more likely to be modern farmers in the sense of having a professional approach to farm work. These two categories reflect changes within agricultural production as well as changes in women's role in society at large. Some statistics are included.

253. Hetland, Per. "Pluriactivity as a Strategy for Employment in Rural Norway." *Sociologia Ruralis* 46:3-4 (1986): 385-395.

Pluriactivity, the traditional employment adaptation in rural Norway, has continued to grow, but the types of job combinations have changed radically from "fisherman farmers" to "white collar farmers." The author uses pluriactivity instead of part-time farming to draw attention to the total activity of the household. Emphasis is also placed on how households are adapting to policies and projects which try to stimulate pluriactivity.

254. Minge-Kalman, Wanda. "Household Economy During the Peasant-to-Worker Transition in the Swiss Alps." *Ethnology* 12 (1978): 83-96.

Looks at certain theoretical propositions about the economics of the transitional peasant-to-worker family among the farmers in the Swiss Alps. Finds that there is a relationship between the level of education and the mother's time allocated to family labor and that the mother's labor hours increase more than the father's as children's educational levels increase. Suggests that the foregone labor of children is a cost to the parents of educating children. This is a hidden cost which has not been included in the growing literature on the cost of children in western industrial societies as compared to less industrial societies.

255. Moerkeberg, Henrik. "Working Conditions of Women Married to Self-Employed Farmers." *Sociologia Ruralis* 18:2-3 (1978): 95-105.

    This study from Denmark analyses the causes of the fact that a growing number of farm wives have been attracted to the labor market as well as the consequences of this development with regard to work in the home and on the farm. A random sample of 320 women was interviewed. It showed that better and more varied employment openings in rural districts as well as structural changes within agriculture have led to an increased demand for jobs. The women came from smaller farms, and they are often young women with vocational training. The changes in women's work situation do not seem to have had any noticeable consequences on the traditional patterns of sex roles in the farm family.

256. Petronoti, Marina. "The Economic Autonomy of Rural Women: A Survey of the Mediterranean with Specific Reference to Three Greek Islands." *The Greek Review of Social Research* 41 (January/April 1981): 6-19.

    The aim of this research is to contrast data on social norms and peasant practices in order to stress the value of comparative analysis in understanding sex roles and the conditions responsible for their definition. Data derive from a number of sources such as an anthropological survey carried out in the Mediterranean as well as Greek folklore studies and reports of local traditions.
    Reviews the literature on Mediterranean women and discusses the values of honor and shame, women's economic activities, female status in Greel law and custom, the reversal of women's domestic position, and gives three cases of Greek female autonomy at the end. The islands included are Kalymnos, Samos, and Karpathos. This is a thesis presented to the University of Kent at Canterbury for the degree of Master of Philosophy in social anthropology.

257. Riegelhaupt, Joyce F. "Saloio Women: An Analysis of Informal and Formal Political and Economic Roles of Portuguese Peasant Women." *Anthropological Quarterly* 40:3 (July 1967): 109-126.

Examines the manner in which the women in a Portuguese village perform economic and political functions within their society. The study took place in the village Sao Joao das Lampas northwest of Lisbon, the administrative center of the parish with a population of 363.

Traditionally this parish has produced wheat, rye, barley, potatoes, cabbages, beans, and carrots as well as grapes for local red and white wine. Most families have a few dairy cattle, chickens, ducks, and turkeys. Men perform the daily and cyclic tasks of agricultural work, and women care for the animals and all the household chores, child rearing, marketing, etc. The Saloio women have contacts outside the village through marketing and by virtue of domestic service; this helps them play informal political and economic roles within their society. The women may function as household and community decision makers, but they will never hold any formal positions, given the present political system. The limits for "women power" are very sharply drawn in Portugal by both the legal principles of the state and the traditional restraints on the role of women.

258. Secretariat of the COPA Women's Committee. *Women in Agriculture*. Brussels: Commission of the European Communities, 1988.

Contains objectives of women farmers in the community and 12 case studies covering the key social aspects which can guarantee promotion of the role of women farmers as well as practical information concerning national women farmer's organizations and organizations promoting equal opportunities for men and women. The countries covered are Germany, Belgium, Denmark, Spain, France, Greece, Ireland, Italy, Luxembourg, the Netherlands, Portugal, and the United Kingdom.

259. Shorter, Frederic C. "The Population of Turkey after the War of Independence." *International Journal of Middle East Studies* 17 (1985): 417-441.

This is a population study rather than a study on agriculture, However, we have included it here, because we do find information on agricultural production and women's participation in the agricultural economy as well as nonagricultural employment of women. Statistical tables are included.

260. Sousi-Roubi, Blanche, and Isabelle von Prondzynski. *Women in Agriculture*. Brussels: Commission of the European Communities, 1983.

Drawn up on the basis of the national reports submitted at the Grado Seminar, Italy, November 1982, this study examines the situation in the ten member states on five legal issues: the different legal situation on the farm, social security, access to vocational training, access to agricultural organizations, and relief services. The member states are Belgium, Denmark, France, the Federal Republic of Germany, Greece, Italy, Ireland, Luxembourg, the Netherlands, and the United Kingdom. Gives also the final report of the Seminar held at Grado, Italy, and reports of inquiry on the problems encountered by self-employed women, particularly in agriculture, trade, and crafts. A list of organizations for women farmers represented in the COPA Women's Committee is also included.

261. Stratigaki, M. "Agricultural Modernization and the Gender Division of Labour. The Case of Heraklion, Greece." *Sociologia Ruralis* 28:4 (1988): 248-262.

Based on a field study in Heraklion, Crete, one of the most agriculturally advanced regions of Greece, this paper examines the transformation of the gender division of labor associated with modernization of agricultural production. In the peasant family farm, most labor is provided by family members. The patriarchal structure of society sets the context for an increasingly discriminatory gender division of labor in farming. Men dominate the mechanized production and the management of the marketing co-operatives, while women are increasingly burdened with agricultural and domestic manual work, and they remain excluded from the co-operatives and other community institutions despite new laws prescribing equality.

262. Sukkary-Stolba, Soheir. "Changing Roles of Women in Egypt's Newly Reclaimed Lands." *Anthropological Quarterly* 58:4 (October 1985): 182-189.

Describes the roles Moslem Egyptian women play in building new communities in the desert land of Egypt. Describes the settlement process of 10,402 families to Tahaddi, northwest of Cairo. Looks at women's motives to migrate, changes in family structure and residence, women's workload, social mobility, and women's participation in community development projects. Life in the new lands offers women settlers many positive opportunities to grow and participate in the public domain. Moreover, settlers overcame the traditional norms restricting female participation in field work, market place, employment, and education.

263. Wood, Lucie Saunders, and Sohair Mehenna. "Unseen Hands: Women's Farm Work in an Egyptian Village." *Anthropological Quarterly* 59 (July 1986): 105-114.

Focuses on changes in women's agricultural work in one village, Tafehnet al Ashraf, located in Daqahliya province in the Delta northeast of Cairo. The period examined extends from the time of Nasser into that of Sadat (early 1960 to late 1970), and the data come from questionnaires, censuses, and interviews conducted in the village. During these years women's participation decreased in agricultural labor and increased in animal husbandry. Finds that the women's participation in production seems to be undercounted in national statistics. The rural women's work in agriculture, animal husbandry, and poultry raising has contributed to the improvement of the material conditions of their lives, to the accumulation of funds for social obligations, and investment of land purchase. The data also show that women's work is sensitive to market factors; thus over time women shifted from agricultural labor towards animal husbandry and poultry production.

GENERAL STUDIES

264. Anker, Richard. "Female Labour Force Participation in Developing Countries: A Critique of Current Definitions and Data Collection Methods." *International Labour Review* 122:6 (1983): 709-723.

Discusses the difficulties involved in obtaining accurate labor force data on Third World women from the point of view of interviewers, respondents, and labor statisticians or economists. Suggests alternative definitions of the labor force and survey questionnaire structure in order to circumvent some of these problems. We also find preliminary results from a field study in order to illustrate these points. Points out that survey designers, economists, and labor statisticians have all too often ignored the difficulties involved in collecting information on the female labor force. Yet, as this article tries to show, respondents cannot be expected to provide accurate labor force data unless interviewers and survey designers ensure that questions relating to labor force activity are very specific. Only by improving field-work techniques and questionnaire design can the statistical invisibility of much of the economic and labor force activity of women be eliminated.

265. Aimed, Iftikhar. "Technology and Rural Women in the Third World." *International Labour Review* 122:4 (July/August 1983): 493-505.

A review of the impact of technological change reveals that changes that accompany "modernization" have, for the most part, led to the concentration of women in domestic and non-market roles and labor-intensive activities. It shows that men take over the responsibilities for women's tasks as soon as they are mechanized or when they are transformed from subsistence into market production.

The first step in helping rural women is to find out which tasks they would like to be relieved of through appropriate technology. It is also important to find out if the women are willing to trade off income-earning opportunities for technological innovations relieving them of the burden and difficulty of work. Predicting the impact of technological change on rural women is complicated by the intricacies of the household/rural labor and production processes and by the lack of appropriate indicators measuring women's welfare.

The author feels since women are not a homogeneous group, policy makers would do well to consider urgently how technology may best be used to benefit certain subsets of women (e.g., female-headed households in Africa and small peasant households in Asia) who constitute the poorest and most disadvantaged of the rural poor.

266. Aimed, Iftikhar, ed. *Technology and Rural Women: Conceptual and Empirical Issues*. London: George Allen & Unwin, 1985.

Is designed to deal with analysis of issues such as gender-based inequalities and the low rate of adoption of improved technologies among women. Brings together conceptual and empirical analysis undertaken by economists, sociologists, and engineers. The book is divided into three parts: conceptual, analytical, and theoretical approaches; two empirical overviews of the impact of technological change on the condition of rural women; and four country case studies from Africa. Suggestions for future research, an extended bibliography, and an index are also included.

267. Beneria, Lourdes, and Gita Sen. "Accumulation, Reproduction, and Women's Role in Economic Development: Boserup Revisited." *Signs: Journal of Women in Culture and Society* 7:2 (1981): 279-298.

When Ester Boserup's *Women's Role in Economic Development* was published in 1970, it represented a comprehensive effort to provide an overview of women's role in development. It became the most cited study and created a flow of research in this area.

Now, more than 10 years later the authors summarize Boserup's main contribution and provide a critical analysis of her approach, particularly in view of recent scholarship on the subject to show the limitations of Boserup's analysis, which, they say arise from a flawed and inadequate conceptual basis. Boserup felt that giving women more education would give them a better chance to compete in urban labor markets and gain access to improved agricultural techniques in the rural areas. The authors feel this conclusion ignores two crucial features that an analysis based on the concept of accumulation and women's role in reproduction would highlight. The authors point out we can no longer ignore the question of what goes on within households nor the interweaving of gender relations and class relations. The feminist analysis of the Third World in the past decade has lent support and clarity to this vision.

268. Beneria, Lourdes. "Conceptualizing the Labor Force: The Underestimation of Women's Economic Activities." *Journal of Development Studies* 17:3 (1981): 10-28.

Looks at the underlying existing labor force statistics and their built-in tendency to underestimate women's contribution to production, particularly in areas with a relatively low degree of market penetration in economic life. Points out the shortcomings of conventional labor force concepts and argues that active labor should include all workers engaged in use value as well as exchange value production, which includes activities such as household production and all type of subsistence production. This can remove the paradox commonly found in the Third World, in which a local economy survives thanks to women's involvement in subsistence production while men are unemployed. Yet official statistics show low labor force participation for women and high participation by men.

269. Blumberg, Rae Lesser. "Rural Women in Development: Veil of Invisibility, World of Work." *International Journal of International Relations* 3 (1979): 447-472.

Formulates some hypotheses concerning female invisibility as well as productivity and contrasts the general view of rural Third World women as unproductive. The evidence is drawn from evolutionary history and data from studies by the United Nations. The former indicates that women were the primary producers in most pre-agrarian human groups, and the latter indicates that women continue to produce approximately half the world's food. Concludes with an analysis of the statistical biases and stereotypes that obscure these contributions and briefly indicates the cost of this invisibility to the countries involved as well as the women themselves. References are included.

270. Charlton, Sue Ellen M. *Women in Third World Development.* Boulder: Westview Press, 1984.

Charlton states in the preface that she seeks to address two audiences: first, the college or university students and teachers who desire an advanced introduction to development means, and, second, the non-university workers who are committed to greater equity in the development process at home and abroad.
The book is divided into three parts: "The Meaning of Development for Women," sets the framework for thinking about development in its historical and political context, "Issues in Research and Public Policy," looks at specific aspects of these processes such as agricultural technology and education, and "Approaches to Development for Women," examines alternative development strategies and the agencies, public and private, that support or fund development projects. Lists of supplementary readings are included after each chapter.

271. Croll, Elisabeth J. "Women in Rural Production and Reproduction in the Soviet Union, China, Cuba, and Tanzania: Socialist Development Experiences." *Signs: Journal of Women in Culture and Society* 7:2 (1981): 361-399.

Examines the degree to which alterations in the relations of production that may be part of socialist development strategies affect the production and reproduction activities of peasant women in four societies: the Soviet Union, China, Cuba, and Tanzania.
First the author looks at how the women in each of these societies have been expected to enter the waged labor force and at the same time continue to service and maintain the household. In the second part of the article the author presents case studies from each of the countries and demonstrates that in the development programs of each of these countries, the emphasis on attracting women into the collective waged labor force has outweighed the concern for redefining women's reproductive and domestic roles.

272. Deere, Carmen Diana. "Rural Women's Subsistence Production in the Capitalist Periphery." *Review of Radical Political Economy* 8:1 (1976):9-15.

Focuses upon one component of rural women's economic participation, subsistence agricultural production, and develops a theoretical framework for the analysis of rural women's contribution to capital accumulation. Shows that the agricultural division of labor in the periphery, with male semi-proletarians and female agriculturalists, contributes to a low value of labor power for peripheral capital accumulation.

273. Dixon, Ruth B. "Women in Agriculture: Counting the Labor Force in Developing Countries." *Population and Development Review* 8:3 (1982): 539-566.

Examines the implications of undercounting women's participation in agricultural activities. Understanding the extent of such undercounting has implications for development planning and theory. Sex-related biases in labor force statistics may lead planners wrongly to assume that women's economic contributions to subsistence of cash crop production, processing, and marketing are negligible. This may result in exclusion of women from access to technical assistance in crop raising or animal husbandry, agriculture credit, training in farm equipment operation, or other resources. Statistical tables and references are included.

274. Ember, Carol R. "The Relative Decline in Women's Contribution to Agriculture with Intensification." *American Anthropologist* 85 (1983): 285-304.

To account for the relative decline in female contribution with the intensification of agriculture, the author suggests that the women are "pulled out" of agriculture because their domestic work increase. She argues that women in intensive agricultural society have more "inside the home" work as compared to women in horticultural societies. Her research also highlights the understanding why it is that, with respect to leadership, the status of women seems to be lower in preindustrial intensive agricultural societies.

275. Flora, Cornelia B. "Women and Agriculture." *Agriculture and Human Values* 2 (Winter 1985): 5-12.

The role of women in agriculture can best be understood through analysis of (1) the relationships of each household member to land through ownership or use right, (2) labor through provision of labor at key times and for key elements in the production cycle, (3) capital, in terms of both the mobilization of inputs and the allocation of the surplus produced. Unlike other types of production, agriculture involves rhythms and risks which influence these relationships.

A major characteristic that distinguishes agricultural production from industrial production is the rhythm of the biological process which determines the time it takes to go from sowing to harvest of crops or from insemination to birth to market for animals. Thus labor time, when labor is actually applied to the production of crops and livestock, is less than production time. Therefore, factors of production must pay attention to irregular production cycle as well as the risks involved due to dependence on nature. Planned programs must keep in mind the productive activities of female farmers, their differential access to land, labor, and capital, and the fact that their productive activities must almost always be combined with their reproductive or household-based activities.

276. Folbre, Nancy. "Cleaning House: New Perspectives on Households and Economic Development." *Journal of Development Economics* 22:1 (1986): 5-40.

Provides a critical review of the resent literature on household and economic development. Folbre moves the household to a more central place within this picture and argues that production for use, production for exchange, and childrearing within the household lie within the purview of economic theory and that more attention to production and distribution within the household could considerably enhance the larger effort to understand and to promote economic development. An extended bibliography is included.

277. Godwin, Deborah D., and Julia Marlowe. "Farm Wives' Labor Force Participation and Earnings." *Rural Sociology* 55:1 (1990): 25-43.

Farm wives' human capital, demands and constraints placed on them by their family and farm responsibilities, and labor market characteristics were examined as influences of wives' off-farm employment and earnings. It was found that the wives' human capital was strongly positive related to their off-farm earnings, while family and farm demands had negative effects on wives' earnings. This study tests, via Tobit analysis, a model which includes all farm wives. The author feels these findings regarding the off-farm labor supply of farm women have important implications. Off-farm employment of wives has become an economic cushion that has permitted many family farms to continue in operation.

278. Godwin, Deborah D., et al. "Wives' Off-Farm Employment, Farm Family Economic Status, and Family Relationships." *Journal of Marriage and Family* 53 (May 1991): 389-402.

Gathers data from 1,067 farm spouses in seven states on different indicators such as income, savings, productive work, and life satisfaction. A multivariate analysis of covariance (MANCOVA) was performed to test the overall effects of wives' off-farm employment on their families' economic and relationship status and functioning. The research shows that there were substantial differences between families with employed and nonemployed wives on all of the personal and family characteristics analyzed. References are included.

279. Gronau, Ruben. "Leisure, Home Production, and Work: The Theory of the Allocation of Time Revisited." *Journal of Political Economy* 8:6 (1977): 1099-1123.

Points out that the household production function is by now an established part of economic theory and that this new consumption theory emphasizes the fact that market goods and services are not themselves the agents which carry utility but are rather inputs in a process that generates commodities, which, in turn, yield utility. An increase in the market wage rate is expected to reduce work at home, while its effect on leisure and work in the market is indeterminate. An increase in income increases leisure, reduces work in the market, and leaves work at home unchanged. These conclusions are supported by empirical tests based on the Michigan Income Dynamics data as well as by previous time budget studies.

280. Heppner, Paul, et al. "An Investigation of Coping Styles and Gender Differences with Farmers in Career Transition." *Journal of Counseling Psychology* 38:2 (1991): 167-174.

Examines the role of career- related variables and coping variables in predicting depressive symptomatology and perceived stress, control, and progress in career transition of a farm population in crisis. The sample of this study consisted of 44 men and 35 women who participated in the Career Options Workshop conducted by the Career Planning and Placement Center at a large Midwestern university. The Problem Solving Inventory and a reused version of the Ways of Coping Scale were used as instruments.

281. Huntington, Suellen. "Issues in Women's Role in Economic Development: Critique and Alternatives." *Journal of Marriage and Family* 37 (November 1975): 1001-1012.

Explores issues and hypotheses concerned with women's work in developing economies. Feels Ester Boserup's unique work *Women's Role in Economic Development* presents a dilemma. Alternative hypotheses, including additional variables such as control of the product of women's labor, extent of work and leisure, domestic work, educational opportunities, etc., explain the dilemma of the African women as well as the differential employment rates of women in modernizing economies. The problems of cross-cultural models of women's work, the economic contributions of women, and the effects of women's work on women's lives are discussed.

282. International Labour Organization. *Women in Rural Development: Critical Issues*. Geneva: International Labour Office, 1980.

Includes a collection of papers presented at a program in Geneva in May 1978 by researchers from the Third World. The purpose was twofold: to establish communication with researchers familiar with the area of rural women in development and to make the ILO's work better known to individuals and institutions concerned with rural women in the Third World. Participants from different regions were asked to make short presentations based on their research and based on the following topics: modes of production and agrarian structures and women's work, sex roles and the division of labor in rural economies, effects on the penetration of the market on rural women, and rural development and women.

283. Jiggins, Janice. "Agricultural Technology: Impact, Issues, and Action" in Gallin, Rita S., et al., ed. *The Women and International Development Annual*. v.1. Boulder, CO: Westview Press, 1989, pp.25-55.

Reviews the changes taking place in the relationship between women and technology and gives directions for future research and policy.

284. Kada, Ryohei. *Part-Time Family Farming: Off-Farm Employment and Farm Adjustment in the United States and Japan*. Tokyo: Center for Academic Publications, Japan, 1980.

A comparative study of part-time family farming in the United States and Japan. Interviews with farm families in both countries were conducted to identify meaningful similarities and differences. Part-time farming is becoming of greater importance as national economic development progresses. Examines the lives of farm families and the influences of farm operation, farm labor, life cycle of the families, aspirations and goals, and off-farm employment opportunities. Each chapter presents relevant statistics and the questionnaires used in the interviews are included, as well as a short bibliography.

285. Lele, Uma. "Women and Structural Transformation." *Economic Development and Cultural Change* 34:2 (January 1986): 195-221.

Summarizes the major themes of the various types of literature relevant to understanding women's role in primary production and economic processes in developing countries. Looks also at the overall trends in production and employment in developing countries over the last two decades before turning to the primary objective, to explore women's distinguishing role as economic actors in traditional societies under quite different social and production organization systems. The author looks at women's influence on production by examining the degree of substitution between their own labor in agricultural production and other activities. Finally, examines policy implication of existing knowledge and identifies areas where further research is critical for improving the understanding of women's role in development. An extended list of references is included.

286. Lewis, Barbara C., ed. *Invisible Farmers: Women and the Crisis in Agriculture*. Washington, D.C.: Office of Women in Development, Agency for International Development, 1981.

Women as primary providers of food locally and nationally have been invisible in development programs. Stresses the gains of integrating women as producers into agricultural development. Women participate in all kinds of cultivation, but they are mostly responsible for cultivating the food consumed in their own communities. When agricultural planners overlook women's labor inputs, they necessarily fail to anticipate crucial in-elasticities in the existing agricultural labor force and to build in incentives for project participation. References are included.

287. McSweeney, Brenda Gael. "Collection and Analysis of Data on Rural Women's Time Use." *Studies in Family Planning* 10:11-12 (1979): 379-383.

This project, one of 37 women-related activities undertaken with the assistance of UNESCO during the decade 1965-1975, was aimed primarily at creating the preconditions for educating women in remote rural areas and at designing education programs that would contribute to rural development. A major objective of the data collection was the generation of precise information on women's time allocation.

Research resources were allocated to a combination of overview and intensive survey techniques, and the data-collection strategy of time budgets was used. Three cross-seasonal time budgets compromising all activities in approximately the first fourteen working hours were prepared for each of the women in the village samples and for women leaders by means of direct observations. Comparison of information on rural women's time use yielded by the recall technique and by direct observation showed that 44% of women's work was unaccounted for.

288. Michaelson, Evelyn Jacobson, and Walter Goldschmidt. "Female Roles and Male Dominance among Peasants." *Southwestern Journal of Anthropology* 23:4 (1977): 330-352.

An analysis of 46 peasant community studies explores feminine roles and ideas. A notable feature of the cultures in the sample is their male centeredness. In 35 of the 46 communities discussed, they found a clear-cut ideology of male dominance. In 26 cases land is inherited in the male line; in 20 cases it is inherited bilaterally, while in no cases is it inherited in the female line. The authors explore the economic setting of this peasant androcentrism and the way in which the masculine bias affects relationships between men and women in peasant households. Statistical tables and references are included.

289. Mickelwait, Donald R., et al., ed. *Women in Rural Development: A Survey of the Roles of Women in Ghana, Lesotho, Nigeria, Bolivia, Paraguay and Peru.* Boulder: Westview Press, 1976.

Survey undertaken for AID by Development Alternatives, Inc., defines guidelines for improving project design when women are considered as a semi-autonomous resource in the rural sector. Gives factors which must be kept in mind by project organizers who seek to effectively mobilize the economic resources that women represent. Includes a country-by-country analysis of data related to the current status of women in the rural sector considering the following factors: legal status, wealth, and inheritance, education, marriage, use of income, opportunities for improving status, etc. Recommendations for further research are included.

290. Mies, Maria. *Indian Women in Subsistence and Agricultural Labour.* Geneva: International Labour Office, 1986.

This study was undertaken under the auspices of the World Employment Program, and the author presents a detailed analysis of working women in three villages in the state of Andhra Pradesh in India. Gives a socioeconomic and historical background of the region and examines the pattern of women's work and the level and sources of their incomes and expenditures. Describes the roles of women as agricultural laborers and their relationship with men as well as the impact of class, caste, and gender on their lives.

291. Norris, Mary E. "The Impact of Development on Women: A Specific Factor Analysis." *Journal of Development Economics* 38:1 (January 1992): 183-201.

Examines the impact of changes commonly associated with the development process on absolute, real, and relative returns of female labor in an agrarian economy. When a specific factor analysis was employed, it showed that the expansion of trade lowers women's absolute, real, and relative returns. Neutral technical progress, when limited to the cash crop sector, has the same qualitative effect on women's incomes. Finally, an exogenous increase in female labor force participation in the cash economy causes also a possible relative decline in the female wages. The author says her findings accord well with stylized accounts of the impact of development on women. We find in the appendix the equations used as well as some figures and a long list of references.

292. Palmer, Ingrid. *The Nemow Case. Case Studies of the Impact of Large Scale Development Projects on Women: A Series for Planners.* West Hartford, CT: Kumarian Press, 1985.

A series of case studies on Women's Role and Gender Differences in Development was developed to demonstrate that such analyses are not only essential but also feasible within existing structures. The author of this case study is evaluating the damming of the Nemow River and the rice development project and their impact on women. Discusses also how attention or inattention to women's roles affected the overall project outcome.

293. Saito, Katherine A., and Daphne Spurling. *Developing Agricultural Extension for Women Farmers*. World Bank Discussion Paper No. 156. Washington, D.C.: The World Bank, 1992.

The World Bank alone has funded 460 projects involving agricultural extension in 79 countries. An Extension Program can increase agricultural productivity and rural incomes by bridging the gap between new technical knowledge and farmers' practices, but researchers and extension services usually assume that farmers are men.

FAO, IFAD and other international agencies estimated that women counted for 70-80 % of household food production in Sub-Saharan Africa, 65% in Asia, and 45% in Latin America. Men's movement into off-farm employment is strengthening women's role as agricultural decision makers, and more and more the farmer in the developing world is a woman.

Since so many women play critical roles in a wide range of agricultural activities, this report provides guidance on how to address the many needs and problems of women farmers and on the design and implementation of agricultural projects for women. The chapters include information on (1) the need to improve extension to women farmers, (2) understanding how gender affects agricultural production, (3) generating appropriate technology for women farmers, (4) improving the delivery of extension to women farmers, (5) guidelines for project preparation design and implementation. Includes references.

294. Sanday, Peggy R. "Toward a Theory of the Status of Women." *American Anthropologist* 75 (1973): 1682-1700.

Examines the factors affecting female contribution to subsistence and the relationship between female production and female status. In developing a theory of the status of women, the emphasis in this paper is on ecological and economic factors. It is clear from the data presented that female production is not simply related to female status. The results indicate that female production is a necessary but not sufficient condition for the development of female status.

295. Schumacher, Ilsa, et al. *Limits to Productivity: Improving Access to Technology and Credit*. International Center for Research on Women, May 1980.

Investigates relations between women and productive resources. Focuses on the nature of definition and extent of women's access, areas of resource needs, and obstacles to both access and use. Why and how resources are integral to women's involvement in the development process, and differences in the situation for men and women are also presented. The roots of sex differences in access to modern productive resources, namely credit and technology, can be due to a combination of factors: economic structure, the way development policies and programs have been introduced by government, international agencies, and private organizations, and also to the limited supply of and demands for modern technology and credit. Because of their lack of economic resources and their reproductive roles, poor women fall more deeply than men into the vicious circle of poverty. The last chapter provides some recommendations for improving women's access to these resources and breaking the circle of poverty. An extended bibliography is included.

296. Sontheimer, Sally, ed. *Women and the Environment: A Reader Crisis and Development in the Third World.* New York: Monthly Review Press, 1988.

Throughout the developing world, ecological equilibrium has been broken by a number of interacting factors, many of which are direct results of development policies geared toward survival in the global economy. These changes in land use and distribution have drastically short-changed the poor, leaving them with not only less access to land but less fertile land. The first essay looks at the critical issue of land, the food crisis and women's role in agriculture, then analyzes how anti-decertification projects have involved women and looks at some of the positive experiences of women's initiatives. The third chapter describes the importance of the forest for women to meet basic needs and provide a source of income; it is followed by a more in-depth study at the cooking energy crisis which is of particular importance for women. We also find an essay on water management and how water supply and sanitation programs have often failed by ignoring women. The last part entitled "Taking Action for a Better Future" includes two studies from Latin America and two studies from India. A few references are included.

SEXUAL DIVISION OF LABOR IN AGRICULTURE

In all farm families there is some division of labor depending on the age and sex of the family members. This division of labor may seem "natural" since it has not changed over a long period of time. But what might seem "natural" to a farm community in one country or on one continent might not be so in another country. In Europe or America, farm women's tasks might be quite different from the responsibilities of the rural women in Africa, India, and Latin America. In this chapter we have included studies related to the division of labor by sex; some authors argue that women's role in reproduction causes the extent and nature of their participation in agricultural production. The fact that women have responsibilities for the children and the daily family maintenance reduces women's mobility and makes the household their primary area of concentration.

Ester Boserup says " an important distinction can be made between two kinds of patterns of subsistence agriculture: one in which food production is taken care of by women, with little help from men, and one where food is produced by the men with relatively little help from the women." She then looks at the division of labour within African agriculture, "the region of female farming par excellence."[1] Boserup's encouragement of further studies in this area released a flood of research from the underdeveloped regions of the world. Money was allocated to introduce modern technology and new ways of agricultural production in the developing world, but little attention was given to the women. The farmer has always been looked upon as male in the developed regions, and the new allocation of funds did not reach the women. They were not included in the statistical information from the regions, and they did not get help or training to improve their agricultural production. Many of the studies included here are pointing out these facts, and that modernization in many cases created an inferior status for women.

The disregarding of the role of women, who are the main producers of food in Africa (more than 60% of the food is produced by women), has revealed another crucial but often-ignored reason why hunger is still prevalent in Africa. Women are invisible farmers; their participation and production is not counted, but they are waking up. Several studies describe the activities of self-help groups being organized in an attempt to open new avenues and overcome the constraints through collective action.

---

[1]Boserup, E. *Women's Role in Economic Development*. London: Allen & Unwin, 1970; New York: St. Martin's Press, 1970.

297. Afshar, Haleh, ed. *Women, Work and Ideology in the Third World*, London: Tavistock Publications, 1985.

Women's work has been severely devalued by a universal ideological framework that regards them as inferior bearers of labor and defines their work as the property of men. This view is strongest in many of the rural areas in the underdeveloped countries. This has been possible by control of the cash economy by men, especially in Islamic countries, where they act as intermediary between women producers and the marketplace where the products are sold. The studies indicate, however, that it is the ideology of male supremacy rather than any specific religion which thus affects women's lives. The case studies prove that it is through childbearing and biological reproduction that the women are recognized, not for the value of their productive labor. Seven of these case studies are related to women working in the fields of agriculture, and three studies refer to industrial employment. Some references, an index, and a list of the participants are included.

298. Barnes, Carolyn. "Differentiation by Sex among Small-Scale Farming Household in Kenya." *Rural Africana* 15-16 (Winter/Spring 1983): 41-63.

Much of the data used for this study were derived from a Division of Labor Module designed by the Central Bureau of Statistics (CBS), Government of Kenya, and administered by women in early 1978. The results are a good general indication of Kenyan small-scale farming households. The CBS defines household as a person or group of persons, generally bound by ties of kinship, who normally reside together under a single roof or several roofs within a compound and are answerable to the same head of the household and share a common source of food. The author gives an historical overview and points out the causes of differentiation among small-scale farmers, social relations of production, factors of production and production activities, and the division of labor.

The data suggest that the husband of the married-female-headed smallholding is employed elsewhere, while the wife maintains the holding, which has usually been inherited or a least claimed by the man. Economic forces alone do not count for the fact that the women maintain the farming household, while men depart. Barnes points more towards a prevailing social meaning of maleness and femaleness that sanctions men leaving the smallholdings and women tending the farm.

299. Bauman, Hermann. "The Division of Work According to Sex in African Hoe Culture." *Africa: Journal of the International Institute of African Languages and Cultures* 1:3 (July 1928): 289-319.

Africa is the land of the most intensive hoe culture. This article gives an overview of the literature on this subject and three tables showing division of labor by tribes: (1) men and women work together at the actual hoe culture; (2) men undertake hoe culture nearly or entirely alone; (3) women alone do field work (except clearing). A long list of references is included.

300. Beneria, Lourdes. "Reproduction, Production and Sexual Division of Labor." *Cambridge Journal of Economics* 3:3 (September 1979): 203-225.

Argues that women's role in reproduction lies at the root of their subordination, the extent and nature of their participation in production, and the division of labor by sex. However, only biological reproduction is linked with women's specific reproductive functions, yet most societies have universally assigned to women two other fundamental aspects of the reproduction of the labor force, childcare and the set of activities associated with daily family maintenance. This reduces women's mobility and makes the household their primary area of concentration. The division of labor by sex in non-domestic reproduction tends to reproduce gender hierarchies at the household level and to create mechanisms of female subordination. Instead of looking at the household as a static unit and the division of labor as "natural" or "given," we must view them as being subject to change and responding to dynamic forces generated by an economy in the process of transformation. A long list of references is included.

301. Beneria, Lourdes, ed. *Women and Development: Sexual Division of Labor in Rural Societies*. New York, N.Y.: Praeger Publishers, 1982.

Illustrates the changing nature of the sexual division of labor as a result of changes taking place in the overall economy. We learn about the integration of rural women into the world market; women workers and the green revolution; the sexual division of labor in the Andes; resource allocation and the sexual division of labor in Nigeria; the creation of a relative surplus population in Mexico; an examination of female Malaysian migrants in Singapore; the impact of land reform on women in Ethiopia; and the sexual division of labor in China. Each chapter has some references, and we find an index and information on the authors in the back of the book.

302. Blood, Robert O., Jr. "The Division of Labor in City and Farm Families." *Marriage and Family Living* 20:2 (1958): 170-174.

A 1954-55 Detroit Area Study interviewed a representative sample of 731 housewives from the Detroit metropolitan area, and 178 farm wives living in three counties west of Detroit were interviewed for the purpose of comparison. The author found a substantial difference in the division of labor by place of residence. It showed that the farm wives carried a larger share of household tasks vis-a-vis their husbands. They also engaged in more home production than the city wives. The author suggests that farm income is relevant. The median urban family income in 1954 was almost double that of the farm families.

303. Bossen, Laurel. "Women in Modernizing Societies." *American Ethnologist* 2 (1975):587-601.

It is commonly assumed that modernization generally brings an increase in sexual equality. This paper examines the alternative possibility, that modernization favors an inferior status for women. The author focus on economic variables as a basis for comparison, and she found by examination of modern changes in various societies in which women's traditional position has been relatively strong, a deterioration in women's position relative to men.

304. Brandth, Berit. "Changing Femininity. The Social Construction of Women Farmers in Norway." *Sociologia Ruralis* 34:2-3 (1994):127-149.

Certain gender-related character traits have always been attributed to women and men. Such characteristics have been connected to the sexual division of labor in society. Since both work and family are in a process of restructuring, old beliefs about gender differences have become untenable. The traditional division of labor between the sexes has changed, since women entered areas of society previously occupied by men, thereby altering what we used to understand as femininity and masculinity.

Examines how such social changes become significant at the personal level. If the condition of being a woman changes, one might also expect gender identity to change. Focuses on women who master heavy agricultural machinery and undertake mechanical tasks on the farm. Studies how the women manage the problem of asserting themselves in a masculine area of work, while at the same time remaining feminine. How do they change, how do they create themselves as women, when they are breaking the gender division of labor by doing the same work as male farmers? List of references is included.

305. Brown, Judith K. "A Note on the Division of Labor by Sex." *American Anthropologist* 72 (1970):1073-1078.

It is suggested here that the degree to which women participate in subsistence activities depends upon the compatibility of the latter with simultaneous child-care responsibilities. Women are most likely to make a substantial contribution when subsistence activities have the following characteristics: the participant is not obliged to be far from home; the tasks are relatively monotonous; and the work can be performed in spite of interruptions.

306. Burfisher, Mary E., and Nadine R. Horenstein. *Sex Roles in the Nigerian Tiv Farm Household*. West Hartford, CT: Kumarian Press, 1985.

Focus is on the division of labor, income, and financial obligations among one ethnic group in Nigeria. Looks at the implications of these divisions for the ability and incentives of each sex to adapt technologies introduced by the agricultural development project. Many charts are included and a short bibliography.

307. Buttel, Frederick H., and Gilbert W. Gillespie, Jr. "The Sexual Division of Farm Household Labor: An Exploratory Study of the Structure of On-Farm and Off-Farm Labor Allocation among Farm Men and Women." *Rural Sociology* 49:2 (1984): 183-209.

Data for this study were gathered by telephone from a random sample of New York State farm households in 1981, a follow-up of a mail survey conducted in 1980. The purpose of the telephone survey was to gather additional data on off-farm labor market participation by members of the farm household. The data show interdependency of the on-farm labor inputs of farm men and women. However, the study indicates a comparable pattern of interdependency with respect to off-farm work. The interdependency of men's and women's on-farm labor is more pronounced among small-farm than among large-farm households.

308. Burton, Michael L., and Karl Reitz. "The Plow, Female Contribution to Agricultural Subsistence and Polygyny: A Log-Linear Analysis." *Behavior Science Research* 16:3-4 (1981): 275-305.

Relationships among plow agriculture, female contributions to crop tending, and polygyny are examined. Without control for world regions, a log-linear analysis suggests that each of these variables is related to the other two. When the authors introduced a control for region with a four-way contingency table, they found significant relationships between region and each of the three variables. Furthermore, they found that the control for region eliminates the relationship between plow agriculture and the female contribution to crop tending. They feel that Boserup's theory that claims that women do less agricultural labor with intensive agriculture is not a valid one.

309. Burton, Michael L., and Douglas R. White. "Sexual Division of Labor in Agriculture." *American Anthropologist* 89 (1984): 568-583.

Examines an ecological explanation for variations in female agriculture contributions and focuses on both of the issues that have motivated concern with the Boserup hypothesis: gender roles and process of agricultural intensification. Formulates and tests a theory of the process of agricultural intensification that explains a high proportion of the variance in female contribution to agriculture. Five variables show replicable effects across two or more regions of the world. These are the number of dry months, importance of domestic animals to subsistence, use of the plow, crop type, and population density. The authors found support for Boserup's idea that there are linkages between agricultural intensification and the division of labor, but they feel their theory is a major modification of Boserup's theory. They introduce two new variables, number of dry months and the importance of domestic animals to subsistence, as much stronger predictors of female participation in agriculture. Statistical tables and a long list of references are included.

310. Connelly, Patricia M., and Martha MacDonald. "Women's Work: Domestic and Wage Labour in a Nova Scotia Community." *Studies in Political Economy* 10 (Winter 1983): 45-72.

Gives a historical overview of the Canadian economy and women's work and then focuses on a fishing community in Nova Scotia with a population of about 300. Analyzes the historical development of the Canadian economy in terms of the changing balance between women's role as domestic labor and as wage labor. Found that women have always constituted an available cheap labor reserve and that women have consistently worked for less than their subsistence, and their domestic labor has meant that total family wages have been less than the true cost of family subsistence. Reveals the heavy burden placed on women both as domestic and wage workers in an economy where full employment and adequate wages for every individual is not possible.

311. Coughenour, Milton C., and Louis Swanson. "Work Status and Occupations of Men and Women in Farm Families and the Structure of Farms." *Rural Sociology* 48:1 (1983): 23-43.

Data from a 1979 survey in Kentucky are used to analyze the relationship between farm acreage and sales to (1) type of farm family as reflected in the farm versus non-farm work status of men and women, (2) the type of occupation of those who work off the farm. It was concluded that the pattern of statues of farm family members has an important impact on the labor process in farm production, and that this varies by type of occupation of the person(s) employed off the farm. The effect of off-farm employment on farm business performance differs for men and women due to differences in their involvement in the farm labor process. Statistical information and references are included.

312. Deere, Carmen Diana. "The Division of Labor by Sex in Agriculture: A Peruvian Case Study." *Economic Development and Cultural Change* 30:4 (1982): 795-811.

Explores the relationship between women's participation within the familial agricultural labor force and the process of agrarian class formation. The focus of the study is the province of Cajamarca, which is the most populous Sierra department of Peru. Census estimates of women's agricultural participation in Latin America uniformly indicate that women's participation in agriculture is insignificant and that it has declined steadily. The author proposes that the female low participation rates are due to faulty conceptual categories for measuring women's agricultural participation, and she illustrates the source of errors and proposes an alternative measure of the extent to which women participate in agriculture. She found that this case study conforms to Boserup's general proposition concerning the relationship between the sexual division of labor and type of cultivation.

313. Deere, Carmen Diana. "Rural Women's Subsistence Production in the Capitalist Periphery." *The Review of Radical Political Economics* 8 (1976): 9-17.

Rural women's subsistence production in the capitalist periphery allows semi-proletarian male workers to sell their labor power to capitalist units of production for less than a subsistence familial wage. Looks at the division of labor by sex based on the articulation between modes of production, which serves to lower the value of labor power for capital, enhancing the relative rate of surplus value for peripheral capital accumulation.

314. Deere, Carmen Diana, and Magdalena Leon de Leal. "Peasant Production, Proletarianization, and the Sexual Division of Labor in the Andes." *Signs: Journal of Women in Culture and Society* 7:2 (1981): 338-360.

Draws on sample data as well as data collected through participant observation in three regions in the Andes. The author says evidence shows that the sexual division of labor among households with smaller farm size reflects the fact that for the majority of smallholders, agricultural production can no longer be the household's principal activity. As agriculture becomes less important, it no longer is the principal occupation of the male head of the household. It is in this sense that the force of rural poverty may contribute toward the breaking down of established sex roles.

315. Dey, Jennie. "Gambian Women: Unequal Partners in Rice Development Projects?" *Journal of Development Studies* 17:3 (April 1981): 109-122.

Argues that agricultural development projects channel inputs to male household heads on the assumption that they control the land, labor, crops, and finance. Research carried out in the Mandinka village of Saruja challenges this assumption, where women cultivate rainfed rice, having ownership or use-rights to rice land, while men control upland and grow groundnut and millet. Three development projects introduced irrigated rice to men who therefore control this land and crop and failed to involve women in rice development schemes. This has increased women's dependence on men and is also a major reason for deficiencies in these projects and low national rice production.

316. Dixon, Ruth. "Land, Labour, and Sex Composition of the Agricultural Labour Force: An International Comparison." *Development and Change* 14:3 (1983):347-372.

Review of census data on agricultural employment in less developed countries reveals clear differences among the sex composition of the agricultural labor force, ranging from 5% to over 50% female. Examines the relationship between the sex composition of the agricultural labor force and selected features of rural economies aggregated at the national level, the distribution of land holdings by size, the market orientation of agricultural production, and the relative attractiveness of urban employment opportunities. Each of these features is shown to exert a sex-specific effect on the farm labor force through the supply side, the demand side, or both. This approach takes nations as units of analysis in an international comparison of the determinants of variations in the sex composition of agricultural workers. This is a different approach in comparison with Ester Boserup's farming system approach. The analysis showed that female share of agricultural labor force in less-developed countries appears highest where the rural economy is characterized by smallholder agriculture oriented toward production for subsistence or for local markets and by low levels of urbanization combined with male-dominant out-migration to towns and cities. Statistical tables and a list of references are included.

317. Dixon-Mueller, Ruth. *Women's Work in the Third World: Agriculture Concepts and Indicators*. Geneva: International Labour Office, 1985.

Addresses a number of methodological issues relating to the conceptualization, collection, and interpretation of indicators of the sexual division of labor in Third World agriculture. Reflects recent efforts to propose a new agenda for creating indicators and improving concepts and methods in the area of women's position in the family and household, education and training, economic production and legal rights, and participation in decision-making in the household and the community. More than 30 tables and figures, a bibliography, and an index are included.

318. Feldstein, Hillary Sims, and Susan V. Poats, eds. *Working Together: Gender Analysis in Agriculture, v. 1 & v. 2*. West Hartford, CT: Kumarian Press, 1989.

Access to resources and effective technologies is often constrained by gender, which can lead to detrimental effects on the organization and implementation of agricultural development programs. Incorporation of gender as an analytical variable in the agricultural development equation is becoming a necessity, and volume one addresses this need by providing the agricultural community with an efficient framework to analyze gender issues in agricultural systems. Volume two provides seven case studies which allow hands-on experience in dealing with gender analysis in research and extension contexts as well as a set of teaching notes for the trainers. The countries included are Botswana, Burkina Faso, Colombia, Indonesia, Kenya, the Philippines, Zambia. Some references are included.

319. Feldstein, Hilary Sims. *Tools for the Field: Methodologies Handbook for Gender Analysis in Agriculture.* Hartford, CT: Kumarian Press, 1994.

Marks the concepts and tools for gender analysis in agriculture. Gender refers to the socially or culturally established roles of women and men and is a social construct. Therefore, women's and men's roles may differ from one place or culture to another and may change over time. The goal of agricultural research and extension is to provide technologies that increase food surpluses and improve farm households' well being. Attention to gender means recognizing that the households in farming systems are made up of women, men, and children who may share, complement, differ, or be in direct conflict in their need for or interest in improved technologies. "Understanding both women's and men's roles gives a richer and more complete picture of a production system," say the editors, Hilary Sims Feldstein and Janice Jiggins in the introduction (page 3). Gender analysis may lead to different research priorities or to different approaches to experimentation and extension than would have happened if only men were considered. The reasons for the use of gender analysis in research are that research becomes more efficient and effective; research may be made more equitable; and the areas where there is the greatest need or opportunity for improved technologies are easily identified and the implications for the whole system seen. The contributions collected in this volume come through numerous channels and are the product of many voices. Covering Latin America, Asia, and Africa, the handbook consists of 39 case studies, illustrating a range of techniques from making gender sensitive interview guides to insuring participatory rural appraisal methods that include gender dimension. It is hoped that these experiences will be examined and adapted by other researchers.

320. Garrett, Patricia. "Women and Agrarian Reform: Chile, 1964-1973." *Sociologia Ruralis* 22:1 (1982): 17-29.

Chile provides one of the few examples of large-scale agrarian reform conducted legally in a non-socialist country during the years 1964-1973. The literature on this experience is extensive, but no analysis considers the implications of the Chilean case for the integration of women into the agrarian reform of rural development process. The author considers the factors which influenced the role of Chilean women in agrarian reform. The principal topics considered are the objectives of reform under Frei (1964-1970) and Allende (1970-1973), provisions of agrarian reform legislation, and changes in reformed units during the Allende administration. The final section considers changes which were and were not possible, that would have enhanced the participation of women in the reformed units and the roles of women as agricultural producers.

321. Gasson, Ruth "Farm Women in Europe: Their Need for Off-Farm Employment." *Sociologia Ruralis* 24:3-4 (1984): 216-227.

Argues that the farm women will experience the same needs as other women for earnings, financial independence, association with others, status and self-fulfillment, but that remoteness, lack of suitable work for women in rural areas, needs of the farm business, and sanctions against farmers' wives working elsewhere may deter them. The balance of motives may change from predominantly financial on small farms to predominantly social and personal on large farms. A review of the literature on off-farm employment of farm women offers some support for these hypotheses.

322. Gasson, Ruth. "Changing Gender Roles: A Workshop Report." *Sociologia Ruralis*, 28:4 (1988): 300-305.

Reports on the 35 papers presented at the theme section on "Changing Gender Roles" at the Bologna Congress in 1988. The papers were of multidisciplinary character with contributions from sociology, economics, anthropology, ethnology, geography, and history. The geographical coverage was worldwide. Points out that most recent analyses of gender relations seek to articulate the productive and reproductive spheres and view only one aspect of women's issues, instead of the much broader problems of women's role within agriculture and society.

323. Guyer, Jane I. "Food, Cocoa, and the Division of Labour by Sex in Two West African Societies." *Comparative Studies in Society and History* 22:3 (1980):335-373.

Examines the cultural context and historical development of the division of labor by sex in the farming system of two tribes of the West African cocoa belt: the Yoruba of Western Nigeria and the Beti of South-Central Cameroon. Attempts to trace the way in which social and cultural differences have been perpetuated by comparing two agricultural economic systems both before and after the integration of the same cash crop. The author found that the present division of labor by sex is best seen as a set of changing constraints rather than a static aspect of social morphology.

324. Hombergh, Heleen van den. *Gender, Environment and Development: A Guide to the Literature.* Amsterdam: International Books, 1993.

Connections between gender, environment and development (GED) have been made since the late 1980s, and this guide looks at gender as it refers to cultural and historical concepts of femininity and masculinity and the power relations between men and women. The term environment refers to natural resources with emphasis on their close relationship with the macroeconomic, political, and cultural environment, and development here refers to the transition from poverty to wealth (economic development). The book is divided into two parts, a listing of introductory texts on various issues related to GED and an extensive bibliography. A list of journals related to the topic and a subject index are also included.

325. Huffman, Wallace E. "The Productive Value of Human Time in U.S. Agriculture." *American Journal of Agricultural Economics* 58:4 (1976): 672-683.

Assesses the quantity and marginal productivity of labor service allocated by husband and wife to their own farm work. A behavioral model of the farm firm is developed and implemented empirically by fitting a production function to county average per-farm data for 1964 for Iowa, North Carolina, and Oklahoma counties.

326. Huffman, Wallace E. "The Value of the Productive Time of Farm Wives: Iowa, North Carolina, and Oklahoma." *American Journal of Agricultural Economics* 58 (1976): 836-841.

Examines data which were gathered from the 1964 Census of Agriculture. The observations were per-farm county average of all 276 counties in Iowa, North Carolina, and Oklahoma, and the productive value of farm wives' time in farm work was assessed by estimating an aggregate production function.

327. International Bank for Reconstruction and Development. *Gender and Poverty in India*. Washington, D.C.: The World Bank, 1991.

Women are vital and productive workers in the Indian economy, yet India invests far less in its female workers than in its male workers. Women also receive a smaller share of what society produces in terms of health care, education, and productive assets that could increase their returns to labor. This report analyzes the current role of women in agriculture, dairying, forestry, and the urban informal sector and offers specific policies and programs for each sector. In addition to broad recommendations about overall approaches to integrating gender issues in India's development planning, the report also suggests specific means by which women can gain wider access to the help, skills, and tools they lack. Many statistical tables and a bibliography are included.

328. Jacoby, Hanan G. "Productivity of Men and Women and Sexual Division of Labor in Peasant Agriculture of the Peruvian Sierra." *Journal of Development Economics* 37:1-2 (1991): 265-287.

Data used for this survey come from the Peruvian Living Standard Survey conducted by the World Bank and the Peruvian Instituto Nacional de Estadistica (INE). Overall, adult male labor was found to contribute more to farm output at the margin than adult female labor, though the extent of the differences is sensitive to how farm output and the labor inputs are measured. Statistical tables and a short list of references are included.

329. Kala, C.V. "Female Participation in Farm Work in Central Kerala." *Sociological Bulletin* 25:2 (September 1976): 185-206.

Deals with the trend of changes in a group of four neighboring villages which come under the reorganized local administrative unit of Trikkateri village Panchayat in the Ottapalam Taluk of Palghat district in India. The villages studied had a population ranging from 1000 to 2000 in 1971. The data was collected through participant observation and interviews with farm laborers. Local data resources permitted a study of the situation in about 1950 and the changes of the 1970s. The author looks at the crop pattern; system of land tenure; the social background of farm labor and the economic setting; and the distribution of work based on sex. The study reveals the trends that have set in from 1950 to 1975 and shows a tradition-bound variation in female roles in local agricultural employment.

330. Keating, Nora, and Brenda Munro. "Farm Women/Farm Work." *Sex Roles* 19:3-4 (1988): 155-168.

A survey of farm women in western Canada showed that while young women were more likely to have a farm work role, they also saw more barriers to farm work than older women. Variations in women's farm work may be a result of cohort differences in socialization for farm work, of farm cycle, or of family cycle. Incorporation of off-farm and household work into the analysis of women's contributions to farm business is suggested as a more comprehensive method of assessing their work involvement.

331. Kohl, Seena B. "Women's Participation in the North American Family Farm." *Women's Studies International Quarterly* 1 (1977): 47-54.

Data for this study were collected as part of a larger study of family life and agricultural enterprise development which has run since 1962. The livelihood of the population is based upon small-scale grain farming and cattle ranching, and the study took place in the sparsely populated region of southwestern Saskatchewan. The intent of the study is to look at the strategic position of women in the small family-owned agricultural enterprise, and the author found that virtually all women were important participants in the ongoing managerial decisions concerning the enterprise.

332. Lancaster, C.S. "Women, Horticulture, and Society in Sub-Saharan Africa." *American Anthropologist* 78 (1976): 539-564.

Reviews the importance of economics in cultural history and looks at women in the tribal political economy; evolutionary theories and African ethnography; prestige economics and the social order; and evolution and the ethnographic present.

333. Lapido, Patricia. "Developing Women's Cooperatives: An Experiment in Rural Nigeria." *The Journal of Development Studies* 17:3 (April 1981): 123-136.

Focuses on the experiences of two groups of Yoruba women who tried to organize themselves along modern cooperative lines. The progress of the first group, which tried to adhere to government regulations, is compared to that of the second group, which molded its own rules. Cohesion, personal development, and financial growth were found to be greater in the self-regulating group. Implications for cooperative policies are also discussed.

334. Lastarria-Cornhiel, Susana. "Female Farmers and Agricultural Production in El Salvador." *Development and Change* 19 (1988): 585-615.

A comparison of female-headed and male-headed farm households in regard to production showed that women farm as well as men even in those cases where they have less access to resources. The analysis presented indicate that women do not have equal access to land in El Salvador, at least within the agrarian reform programs, and few women are cooperative members. The author found that most programs in Latin America directly benefit male-headed households.

335. Leacock, Eleanor, and Helen I. Safa. *Women's Work: Development and the Division of Labor by Gender*. South Hadley, MA: Bergin & Garvey Publishers Inc., 1986.

Based on a conference on the Sexual Division of Labor, Development, and Women's Status which took place in Austria, this book yields new insights into important questions regarding development policies and women's status today. The content deals with the issue of gender inequality both in the private and the public domain. The gender division of labor at the household level and at the societal level demonstrates that no strategy of change confined to one level alone can succeed in eradicating female subordination, even under socialism. An evolutionary perspective on the gender division of labor enables one to see that the relation between the domestic and public domains is not static but is highly responsive to changes in the relations of production as a whole. Eight of the chapters in the book have been published in a special issue of *Signs* 7:2 (1981). An extensive list of references and an index are included.

336. Leckie, Gloria J. "Female Farmers in Canada and the Gender Relations of a Restructuring Agricultural System." *The Canadian Geographer* 37:3 (1993): 212-230.

Provides a brief overview of female farm operators and their farms in Canada, using Agriculture and Population data from 1971, 1981, and 1986. Examines the socially constructed daily life of female farmers, using interviews with a sample of female farmer operators in Ontario. Looks, in particular, at the reasons these women chose this occupation and tries to identify the mechanisms by which the traditional gender relations of farming are maintained and reproduced. Central to this are the constraint of women's access to key agricultural resources through issues related to inheritance, the perceived legitimacy of women in this nontraditional role, and the ongoing processes of agrarian myth making.

337. Loutfi, Martha Fetherolf. *Rural Women: Unequal Partners in Development*. Geneva: International Labour Office, 1980.

Draws together the principal themes of rural women's work. The perception of status by rural women themselves and by others is then reviewed and discussed in relation to the women's work. The effects of official policies on rural women are highlighted. The urgency of providing opportunities for increased cash earnings along with reducing the work burden is stressed, and the dependence of effective policies on the participation of rural women is explained. A selected bibliography is included.

338. Long, Norman, ed. *Family and Work in Rural Societies: Perspectives on Non-Wage Labour*. London: Tavistock Publications, 1984.

Papers presented at a Working Group on Non-wage Remuneration and Informal Co-operation in Rural Society held in Espoo, Finland, in 1981. The chapters are grouped around three principal themes: the factors that promote or inhibit peasant forms of organizations and the family farm; the question of inter-household exchanges and community-level patterns of cooperation; and the analysis of the changing roles of women within the context of the household division of labor. These three themes are closely interrelated and of central importance in the analysis of the social consequences of capitalist expansion among agrarian populations, both in Europe and the less-developed countries. A long list of references, a name index, and a subject index are included.

339. Mandala, Elias. "Peasant Cotton Agriculture, Gender, and Inter-Generational Relationships: The Lower Tchiri (Shire) Valley of Malawi, 1906-1940." *African Studies Review* 25:2-3 (1982): 26-44.

Describes the effects of the emergence of a cash-crop economy on rural production relations, particularly gender and inter-generational relationships. The channeling of both male and female labor into cotton production played a significant role in the process by which both sexes redefined their positions in society. The availability of cash-earning opportunities for women allowed them to realize the value of their labor in a more direct manner than had be possible before. Ethnicity, which played a significant role in pre-colonial political relationships, was not very important in the transformations of gender and inter-generational relationships. There were two main causes for the collapse of the cotton economy; (1) the rising level of the waters of the Tchiri River and Lake Malawi flooded large areas of the valley, and (2) the Great Depression plummeted cotton prices, and buyers refused to purchase lower grades of cotton; this hit the producers severely, since much of their cotton was lower grade.

An estimate of two thousand men left the district as migrant workers, and altered the ratios of female-headed households and male-headed households at a time when opportunities for women to earn cash at home diminished. A list of references is included.

340. Massiah, Joycelin, ed. *Women in Developing Economies: Making Visible the Invisible*. Providence, RI: Berg Publishers Inc., 1993.

Selection of studies and articles aimed to sensitize planners and decision-makers to the invisible socio-economic and cultural contribution of women in developing countries is presented.

The purpose is to collect information, analyze its contribution to national economies, and give women adequate financial support or training.

In order to attain sustainable development, the often-forgotten social and cultural dimensions have to be systematically integrated into development. How can we make the contribution of women visible and more productive? These are questions that planners and decision-makers will have to face in the future, and this publication will further this debate. Includes studies from the Caribbean, Tunisia, Ghana, Colombia, and the final chapter looks at integrating women in development planning in India. Each chapter includes a long list of references.

341. Mackenzie, Fiona. "Local Initiatives and National Policy: Gender and Agricultural Change in Murang'a District, Kenya." *Canadian Journal of African Studies* 20:3 (1986): 377-401.

Focuses on three specific areas of change and places particular women's action in the historical context of agricultural policy. Argues that the recent locally initiated changes in Murang'a District are direct response to socio-economic structural change arising from policies initiated during the colonial area and pursued by the post-Independence government with its "national development" strategy. A recent random sample study from 300 households in the Murang'a District indicates women are solely responsible for the provision of food. Women's insecurity has increased as a result of the individualization of land tenure under increasing unequal distribution of this resource.

342. Mayoux, Linda. "Integration Is Not Enough: Gender Inequality and Empowerment in Nicaraguan Agricultural Co-operatives." *Development Policy Review* 11:1 (March 1993): 67-89.

Discusses the experience of women in agricultural co-operatives in Nicaragua under the Sandinistas, based on research in Matagalpa and Esteli Regions at the end of 1988. Co-operatives have often been promoted as the ideal type of project for women, combining possibilities for both income-earning and consciousness-raising. For many of the women interviewed, child-care was seen as the most severe restraint on their participation in the co-operatives and also in training and management. Even where women did not have the problem of young children to care for, their ability to take part in production or other co-operative activities was severely restricted by their responsibility for unpaid and time-consuming domestic work, such as food processing and cooking, house cleaning, fetching water, and washing clothes. The amount of domestic work women had to do themselves, rather than sharing it with other women, children, or adult males, depended very much on their marital status, class position, and the number, age, and sex of their children. However, unpaid domestic work was seen for all of them as their responsibility and their main obligation even when they were involved in agricultural production. Failure to address these problems was not simply a question of inadequate resources. There was considerable institutional resistance to feminist issues. In the context of the agricultural co-operatives studied, addressing the issue of reproductive work did not necessarily require significant amounts of additional resources. Addressing the wider issues of gender inequalities was not a question of resources but of political commitment and organization. A list of references is included.

343. Montgomerie, Deborah. "Men's Job and Women's Work: The New Zealand Women's Land Service in World War II." *Agricultural History* 63:3 (Summer 1989): 3-13.

Efforts to extend the use of female labor in the farming sector provide an interesting example of the obstacles which confronted wartime attempts to dispense with or modify definitions of gender-appropriate behavior in the paid work force. An examination of the difficulties experienced by the government-sponsored Women's Land Service highlights the persistence of conservative stereotypes of women workers, government reluctance to challenge these stereotypes, and the resilience of the sexual division of labor in wartime.

344. Moock, Joyce Lewinger, ed. *Understanding Africa's Rural Households and Farming Systems.* Boulder CO: Westview Press, 1986.

Examines the relationship between farming system research and studies of household decision-making as each bear upon the improvement of food production in Africa. Questions are now being asked about whether current agricultural policies, extension services, and technology transfers are capable of creating surplus among the household-based cropping and herding communities that form the mainstay of African agriculture. The papers were presented at a conference in Bellagio, Italy, in 1984 to assess the state of knowledge regarding household decision-making dynamics in Africa and to explain how these processes relate to farming system research and agricultural policy. A number of papers accuse farming system research (FSR) of paying insufficient attention to the multiple objectives of individual farmers within the household. The complex linkage between intra-and inter-household variables in determining resource access and control was another theme to emerge as well as gender issues in farming systems research and agricultural policy. A bibliography and a list of contributors are included.

345. Moore, Keith M. "Agrarian or Non-Agrarian Identities of Farm Spouses." *Rural Sociology* 54:1 (1989): 74-82.

Explores the relationship between the identification of farm husbands and farm wives with agrarian values and related sex role orientation and position in social structure of agriculture. The data for this study were based on a sample of Wisconsin farm couples (179 usable cases) interviewed in 1979. New identities and values are emerging as farm spouses recognize the decline of traditional family farming in their own lives. Two aspects are significant in this regard: the shift from identification with production to identification with lifestyles not directly dependent on agriculture and the declining subordination of farm wives due to their increased off-farm employment.

346. Mwaniki, Nyaga. "Against Many Odds: The Dilemmas of Women's Self-Help Groups in Mbeere, Kenya." *Africa* 56:2 (1986): 210-228.

A growing body of literature focusing on women's contribution in the development process has revealed another crucial, but often ignored, reason why hunger is still prevalent in Africa, namely the disregard of the role of women, who are the main food producers in Africa. This study concerns some women in Mbeere, Kenya, who have organized themselves into self-help groups in an attempt to open up new avenues and overcome the constrains through collective actions. Describes the activities these groups initiated to raise the capital needed to launch income-generating projects and overcome the labor crisis. Also discusses the problems that inhibit the group's activities and concludes with an evaluation of the feasibility of the proposed income-generating projects. A long list of references is included.

347. Pankhurst, Donna. "Constraints and Incentives in 'Successful' Zimbabwean Peasant Agriculture: The Interaction between Gender and Class." *Journal of Southern African Studies* 17:4 (December 1991): 611-632.

Analyzes the context of social differentiation in which agricultural production takes place in rural Zimbabwe. Crops are grown and livestock raised under conditions partly determined by divisions in village life along lines of gender and class - a fact rarely noticed by policy analysts. A village case study highlights the ways in which production and reproduction overlap in reality, if not conceptually. Offers also insights into the interaction between gender and class dynamics which have implications for southern African theoretical debate and points out implications for future communal area policy options.

348. Pearson, Jessica. "Note on Female Farmers." *Rural Sociology* 44:1 (1979): 189-200.

Informal interviews of 11 female farmers give an exploratory but accurate impression of female farming roles in one county in Colorado. The author says we need to know how female production roles vary in different geographical and technological settings. The size of a farming operation, its degree of mechanization and specialization, the types of crops grown, and the commercialization and wealth of a farm, may all feature critically in determining farm labor requirements and the production roles played by women, and a national survey is required for such analysis.

349. Peters, Pauline. "Gender, Developmental Cycles and Historical Process: A Critique of Recent Research on Women in Botswana." *Journal of Southern African Studies* 10:1 (October 1983): 100-122.

Selectively raises a number of critical questions about present trends as they relate mainly to rural women in Botswana. The particular challenge has been to integrate macro and micro perspectives. First the author illustrates the shortcomings of an undifferentiated perspective on the labor migration economies of southern Africa, with particular reference to Botswana and Lesotho; then she argues that patriarchy is not a useful answer to the model's particular failure to consider gender. She considers the achievements and disadvantages of the close focus on household topologies, and examines current understanding of female-headed and matrifocal households.

350. Pfeffer, Max J. "The Feminization of Production on Part-Time Farms in the Federal Republic of Germany." *Rural Sociology* 54:1 (1989): 60-73.

Nationwide data from the Federal Republic of Germany indicate that about 40% of all farmers are part time farmers. Evaluates the argument that women on part-time farms become responsible for a wider range of farm tasks than the women on full-time farms and that these women are responsible for the continued operation of the part-time farm. The process by which women have taken responsibility of running the farms is referred to as feminization of production. The analysis showed that labor inputs on part-time farms were lower than on full-time farms, but it pointed to a proportionately more extensive involvement of wives on part-time farms. The feminization of production on part-time farms is not the result of women taking over male-dominated tasks, but it represents greater involvement of women in those tasks in which they previously had more limited involvement. Women tend not to be employed off the farm in Germany, and the viability of part-time farms depends on the continued willingness of women to do farm work.

351. Poats, Susan V., et al., Eds. *Gender Issues in Farming Systems Research and Extension*. Boulder: Westview Press, 1988.

Papers presented at an international conference hosted by the Women in Agricultural Development (WIAD) Program at the University of Florida in 1986. Seeks to support and develop expertise related to the roles of women and intra-household dynamics in agricultural production, research, and extension. Divided into four parts, the book first covers topics related to theoretical and methodological implications of the inclusion of gender in farming system research and extension. It then discusses these topics in relation to Latin America and the Caribbean, Asia and the Middle East, and Africa. A list of the contributors is available.

352. Radcliffe, Sarah A. "The Role of Gender in Peasant Migration: Conceptual Issues from the Peruvian Andes." *Review of Radical Political Economics* 23:3-4 (1991): 129-147.

Examines the fundamental role played by gender relations in generating and shaping characteristics of peasant migration by focusing on a case study from Andean Peru. Argues that gender is a major conceptual tool for understanding the organization of peasant migration, as peasants organize participation in subsistence-oriented agricultural production and in labor/product markets via gender criteria in labor allocation. Analysis of the construction of gender relations and household peasant production provides insight into why women leave at a younger age than men, make more permanent moves to cities, and usually terminate migration upon marriage.

353. Reimer, Bill. "Women as Farm Labor." *Rural Sociology* 5:2 (1986): 145-155.

Argues that both the orthodox and radical treatments of farm labor are distorted by the under-representation of women's participation on the family labor farm. Presents data that demonstrate the unity of farm and household organized around three claims: (1) contribution of the family labor farm is underestimated by official statistics; (2) the narrow definition of agricultural labor used by the orthodox and radical theorists excludes women more than men; and (3) the activities included in the broader definition of agricultural labor make a significant contribution to the operation of the farm. Variables examined include the characteristics of both paid and nonpaid laborers, the distribution of responsibility for household and farm tasks, the time spent in various activities, and the extent to which household goods are produced by family members. The result indicates that both the direct and indirect contribution of women must be integrated into the analysis of agricultural production. Statistical tables and some references are included.

354. Rothstein, Frances. "Women and Men in the Family Economy: An Analysis of the Relations Between Sexes in Three Peasant Communities." *Anthropological Quarterly* 56:1 (1983): 10-23.

Data for this study come from the author's fieldwork in the early 1970s by the author in one central Mexican community, San Cosme, and information collected previously by Oscar Lewis and Ralph Beals in the 1940s on two other central Mexican communities, Tepoztlan and Cheran. The author wants to prove that the variations shown in male-female relations between these two studies are related to the approaches used in description and analysis. After a brief description of the communities, the author examines the roles of peasant women and men and summarizes the findings which suggest that the differences in male-female relations are more apparent than real.

355. Sachs, Carolyn, E. "American Farm Women." in *Women and Work: An Annual Review.* Vol. no. 2. Stromberg, Ann H., et al., eds. Beverly Hills. California: Sage Publications, 1986. p. 233-248.

An overview of women's agricultural activities in the United States shows that the sexual division of labor on farms continuously changes with shifts in the economic structure in agriculture. Overall, the major shift over the last decade has been toward a decline in numbers of both men and women involved in agriculture. Today women work on family farms as well as farms not owned by their families, and the author found that women's labor on nonfamily farms is increasing compared to that of men.

356. Sagrario Floro, Maria. "Market Orientation and the Reconstitution of Women's Role in Philippine Agriculture." *Review of Radical Political Economics* 23:3-4 (1991): 106-128.

Examines how agricultural commercialization, as a result of export cropping, has affected women both as workers and as family members. Data on the time allocation of 374 women in the Philippines allow investigation of the changes brought about by the shift from corn (semi-subsistence) farming to sugar (export) production on the magnitude and form of women's productive roles at home, on the farm, and in the market. The data demonstrate how a crop shift alters the parameters of women's time allocation decisions and thereby reconstitutes women's productive roles. This dimension is largely ignored in policy impact studies because women's economic contribution is not recognized. Any benefits or costs in terms of women's work are not taken into account in any evaluation of trade or growth strategies, less so in any formulation of structural adjustment plans.

357. Saito, Kathrine A. "Extending Help to Women Farmers in LDCs: What Works and Why." *Finance and Development* (September 1991):29-31.

Summarizes the reasons why women farmers in Africa need help. In much of the world, land title is in the name of the male head of the household, and women are allocated lands that are far from their village and far from other plots they cultivate. This pose problems of childcare and transportation. Lack of suitable farm and household technology also impairs women's efficiency, and in most of the developing world women are bypassed by formal credit systems because of lack of collateral (usually land title). The women also have less time and mobility, in most countries cultural norms affect interactions between agents and farmers, and finally women are less educated than men, which prevents them from understanding and using technical information. Accurate information is important at the local level (what women farmers do, how they do it, and why they do it) if the research and extension services are to produce and deliver technical advice that is both needed by the farmers and appropriate to their circumstances. It is also better to use women extension agents in training these women, and for that it is important to boost the number of women going into agricultural education.

358. Salamon, Sony, and Ann Mackey Keim. "Land Ownership and Women's Power in a Midwestern Farming Community." *Journal of Marriage and Family* 41:1 (1979):109-119.

Data obtained from interviews with and observation of 22 households of German extraction in central Illinois showed control of land, a scarce resource for farmers, is found to be the source of women's power in this community. Because men control the actual farming and distribution of what is produced, women generally relinquish power obtained by means of land ownership to husbands or male relatives. This shift of authority and other agricultural practices in rural America is seen to account for male domination in both family and community. Women appear to make a trade-off of lower status and less power for male management of the family enterprise which assures them a financially secure widowhood. Some statistical tables and a few references are included.

359. Sheehan, Nancy. *Workshop Proceedings for Gender and Natural Resource Tenure Research.* University of Wisconsin-Madison, Land Tenure Center, 1992.

The workshop was to meet three objectives: to explore the question of how the social relations between men and women determine the multitude of tenure regimes and land-use patterns around the world; to provide information on a variety of research methods for assessing and understanding resource use and tenure patterns resulting from these social relations; and to discuss theoretical and methodological issues in terms of the Land Tenure Center (LTC) research mandates, that is, to identify present research gaps and future research priorities. An account of the presentations and discussions during the workshop is included as well as the case studies.

360. Simpson, Ida Harper, et al. "The Sexual Division of Farm Household Labor: A Replication and Extension." *Rural Sociology* 53:2 (1988): 145-165.

A study of how New York State farm couples allocated their labor across on- and off-farm work is replicated and extended. Farm, family, and individual characteristics influenced the differentiation and integration of husbands' and wives' on- and off-farm work in much the same manner as revealed in the replicated study, but these new data show that the effects of these factors are mediated by crop. The authors feel that future research should be careful to take into account the combined effects of farm size, production system, and how the labor supply available is distributed across gender and age categories. Several statistical tables are included.

361. Spring, Anita. *Women Farmers and Food in Africa: Some Considerations and Solutions.* East Lansing, Michigan: Women in International Development, Working Paper #39, Michigan State University, 1987.

Women farmers in Africa do not constitute a homogenous group. Some are modern growers using scientific methods and commercial procedures; most are not very efficient and are constrained by the lack of knowledge and access to resources as well as by labor burdens. The author urges the development planners to consider the local conditions, and they must target women as farmers, sellers, processors, and preparers of food in order to better their conditions.

362. Stratigaki, Maria. "Agricultural Modernization and Gender Division of Labour: The Case of Heraklion, Greece." *Sociologia Ruralis* 28:4 (1988): 248-262.

Examines the transformation of the gender division of labor associated with the modernization of agricultural production in Heraklion, Crete, one of the most advanced agricultural regions in Greece. In the peasant family farm, most labor is provided by family members. The patriarchal structure of the Greek society sets the context for an increasingly discriminatory gender division of labor in farming. Men dominate mechanized production and the management of the marketing co-operatives, while women are increasingly burdened with agricultural and domestic manual work, and they remain excluded from the co-operatives and other community institutions despite new laws prescribing equality.

363. Straus, Murray A. "Family Role Differentiation and Technological Change in Farming." *Rural Sociology* 25 (June 1960): 219-228.

To test the hypothesis that farm operator technological competence is associated with an "integrative-supportive" wife marital role, a sample of

903 Wisconsin farm operators was classified into matched high and low technology competence groups. The wives of these two groups were compared on 46 items hypothesized to reflect this concept. One third significantly differentiated the two groups. The differentiating items are used to form a Wife Role Supportiveness Index. Multiple and partial correlation shows this index to add a significant increment to the total explained variance in use of improved farm practices. Implications for theory and action programs are also discussed.

364. Tigges, Leann M., and Rachel A. Rosenfeld. "Independent Farming: Correlates and Consequences for Women and Men." *Rural Sociology* 52:3 (1987): 345-364.

Independent farmer is defined as "a woman or man with managerial responsibilities for the farm operation who does not have a spouse regularly performing farm labor." Men without the direct farm labor of a wife did not appear worse off economically than men who had this help. Independent women farmers were worse off than women on other farms. Class position explained only some of the differences among women. Family and demographic characteristics explained the remaining effect of women's independent farming.

365. Tinker, Irene, ed. *Persistent Inequalities: Women and World Development.* Oxford: Oxford University Press, 1990.

Dedicated to Ester Boserup and her pathbreaking work, *Women's Role in Economic Development*, as well as to all the women who lobbied for and during the United Nations Decade for Women to put women on the world agenda. Offers an overview of the past and current debates in the field and shows the connection between the field and the reality of women's work and lives within and outside the household. Shows also the global women's movement and the impact that these advocates and new scholarships on women have had on the policy and programs of development agencies. Most studies of women in development have been undertaken by development agencies and tend to exist in fugitive form, and this research has not yet been widely utilized by the academic community despite its value as real world case studies. On the other hand policy planners now design programs to ensure that women participate in and benefit from development programs, but new agricultural crops and new technologies have increased women's work burden as often as they have reduced it.

366. Tinker, Irene, and Michele Bo Bramsen, eds. *Women and World Development*. American Association for the Advancement of Science: Overseas Development Council, 1976.

A seminar took place in Mexico City in June 1975 to show how and why development programs often have failed to reach women and to emphasize the waste of human potential that has resulted from ignoring the contribution of women to economic and social growth. It also sought to bring together researchers and planners of development programs in order to bridge the gap. Five simultaneous workshops concentrated on these subjects:
(1) food production and the introduction of small-scale technology, (2) urban living, migration, and employment, (3) education and communication, (4) health, nutrition, and family planning, (5) women's formal and informal organizations. A preliminary bibliography of published and unpublished works on women in development was also prepared prior to the seminar by Mayra Buvinic. This was later published as an expanded annotated bibliography (see entry under bibliographies).

367. Toth, James. "Pride, Purdah, or Paychecks: What Maintains the Gender Division of Labor in Rural Egypt?" *International Journal of Middle East Studies* 23:2 (May 1991): 213-236.

Questions just how rigid rural Egypt's gender division of labor is and what maintains the distinction of women's work from men's work. Toth begins by examining the gender division of labor in the Middle East and then focuses on how Egyptian farm work is genderized. He also discusses why such divisions continue and why, in the last quarter of the twentieth century, the supposedly clear lines of separation have become so confused. However the gender division of labor in Egypt has generated more than local problems. In 1961 it spelled disaster for plans of industrialization; in the 1980s it has reduced Egypt's food self-sufficiency. Rural Egyptian women have clearly contributed significant amounts of labor to producing food and exports. When economic diversification permitted unskilled male farm workers to opt out of agricultural employment into migrant labor, construction jobs, and even overseas work, women replaced them, though not without the low esteem and the low wages such employment generates. Resistance to this manipulation has resulted in serious production mishaps.

368. Wallace, Tina, and Candida March, eds. *Changing Perceptions: Writings on Gender and Development.* Oxford: Oxfam, 1991.

Analyzes the global crises on women and why some of these international issues specifically affect women and women's relation with men. Studies of the impact of debt or famine, for example, focused on men and ignored the changes imposed on women's lives and the ways in which relations between women and men are changing under these external pressures. This lack of understanding of the women's situation has led to inappropriate and even damaging development or aid programs. The analyses presented focus predominantly on women living in poverty. Looks at barriers to women's development and talks about how aid agencies now accept the crucial role women play and the many real needs they have and focuses on ways in which agencies can develop their understanding of the needs of women. Appropriate methodologies for planning and evaluating projects and programs for their impact on women must be developed. The book gives several case studies and in the final chapter discusses gender debates and how an individual's reactions to gender issues are rooted in experiences of and attitudes towards power, politics, culture, and everyday social interaction. Some references are listed after each chapter, and an index is included.

369. Whatmore, Sarah. *Farming Women: Gender, Work and Family Enterprise.* London: Macmillan Academic and Professional Ltd., 1991.

Placing women at the center of analysis, this book challenges the prevailing invisibility of women in farming and family enterprise and portrays their distinctive experiences as "farm wives." Through a detailed study of family farming in England, the political economy of family-based production is examined as a unity of household and enterprise, intimately structured by patriarchal gender relations. The research for this book took place between 1985 and 1988, and the central concern is the reconstruction of the political economy of family-based production from a feminist point of view. "The book provides a detailed example of the way in which class relations and forms of production are partly constituted out of the opportunities for authoritative power created by patriarchal gender relations," says the author in the conclusion (page 146). A bibliography and an index are available in the back.

370. Whatmore, Sarah. "From Women's Role to Gender Relations: Developing Perspectives in the Analysis of Farm Women." *Sociologia Ruralis* 28:4 (1988): 239-247.

The guest editor of this issue of *Sociologia Ruralis* focuses on the position of women in family farming systems and examines the development and contribution of feminist theory and research in relation to the analysis of the position of women in farming. Argues that the study of farm women forces us to look more critically at some of the most basic but often taken-for-granted concepts associated with the analysis of the farm labor process. Attention is focused on three areas within feminist work: the nature of gender as a social relation, the family, the labor process.

371. White, Douglas R., et al. "Sexual Division of Labor in African Agriculture: A Network Autocorrelation Analysis." *American Anthropologist* 83 (1981): 824-849.

A model of causes and consequences of sexual division of labor in agriculture is tested using as sample some African societies. Autocorrelation analysis of relationships among crop type, slavery, residence, polygyny, and female participation in agriculture has validated the authors' model. Crop type and the presence or absence of slavery are shown to be effective predictors of the degree of female contribution to agricultural subsistence, and the degree of polygyny is shown to be affected by female agricultural contribution and the form of residence. This effect is an example of one of the phenomena that anthropologists have referred to as Gallon's problem. Many tables and figures as well as a list of references are included.

372. Wilson, John. "Public Work and Social Participation: The Case of Farm Women." *The Sociological Quarterly* 31:1 (1990): 107-121.

The impact of labor force participation by women on social activism is examined using data gathered in a study of 695 North Carolina farm families. The traditional gender divisions of labor break down when women take off-farm jobs. Their rates of activism in instrumental organizations move closer to those of men. Labor force participation by women also reduces the impact of the husband on the wife's activism.

373. Young Kate. "Modes of Appropriation and the Sexual Division of Labour: A Case Study from Oaxaca, Mexico" in *Feminism and Materialism: Women and Modes of Production*, Kuhn, Annette, and AnnMarie Wolpe, eds. London: Routledge and Kegan Paul Ltd., 1978, pp.124-154.

Compares the position of women over a relatively short period of time, a period that spans crucial changes in modes of production, by examining the specific conditions of sexual division of labor as affected by merchants and circulation capital. Deals also with the effect of changes of land usage on the sexual division of labor. The author notes that the sexual division of labor of the traditional society has begun to change. Today socioeconomic status is as important a variable as sex in determining the division of labor. While the rich women may run a family business, perhaps remaining in the shop while the man goes out to oversee the hired laborers in the fields, a woman should not be an independent trader and thus socially visible. The women with less income have often to take on income earning employment, but within the home, which brought a lower return than other, more visible, work outside the home.

374. Young, Kate. *Planning Development with Women: Making a World of Difference*. New York: St. Martin's Press, 1993.

Intended for students in the social sciences considering working in some aspects of development. Gives an overview of the terms and evolution of the main debates from 1960 to 1990 and explains how women fit into these debates. Two chapters review changes in development strategies during this time, touching on the impact of policies and strategies on women. Suggests how development planning can be adapted to support and enhance women's participation and empowerment. Statistical data collection is also discussed, and two aspects of this complex issue are examined: women's economic activity and the household as a unit of analysis. The need for statisticians, planners, and economists to re-examine their theoretical frameworks is argued, and basic concepts like productive and reproductive work should be re-evaluated. Some of the issues involved in planning development with women are also examined. An extended bibliography and a list of further reading are included as well as an index.

375. Young, Kate, et al., eds. *Marriage and the Market: Women's Subordination in International Perspective*. London: CSE Books, 1981.

A collection of articles by women who have been active in the women's movement in their own countries or who have taken part in the debates raised by the feminist challenge to the orthodox interpretations of women's position in society. A Subordination of Women Workshop took place in 1981, and several of the articles included here are from the discussion there. The editors believe there are many fundamental causes of gender subordination, and in this volume they want to identify the diverse elements and mechanisms of women's continuing subordination. Topics such as gender and economics, subsistence production and reproduction, households as natural units, authority within the household, politics of domestic budgeting, kinship and capitalist development, sexuality and the control of procreation, and women in socialist societies are included. A bibliography and a list of notes on the contributors are also available.

DECISION-MAKING ON THE FARM

A long-term case study at Cornell University found that agricultural choices are flexible and responsive to changes in the farm family decision-making environment. Farmers are influenced by needs and goals as well as the resources available to them. The expansion of and inclusion of women in agricultural higher education made a difference in farming families' decisions to include daughters and daughters-in-law in ownership and management of the farms. A study by Susan Rogers uses a model to explain the myth of male dominance, that a balance is actually maintained between the informal power of women and the overt power wielded by men.

Other studies found that decisions were related to women's information-seeking activities and their involvement in farm tasks and family income. Eugene Wilkening, who has done by far the most research in this area, measured the dimensions of aspirations, allocation of tasks, and decision making and found interrelationships among these dimensions. One study showed that the wife's aspirations are more highly associated with mechanization than those of the husband, that she views labor-saving equipment as more of an improvement than her husband, and that the adoption of improved farm practice is higher when both husband and wife have high aspirations for farm improvement. A study from central Illinois by Salamon and Keim on land ownership and women's power demonstrates that the ability to use power in the community depends directly or indirectly on control of land. They point out "that the female control of land must be recognized by those who work with farm families and those who formulate public policies."[1]

376. Abell, Helen C. "Decision-Making on the Farm." *The Economic Annalist* 31 (February 1961): 7-9.

A study describing typical Ontario farm homes and homemakers. The families on these farms are a close knit social unit where both the husband and wife, and even the children, help make decisions regarding the farm. A total of 352 farm homes were visited, and it was found that in most of the families

---

[1]Salamon, Sonya, and Ann Mackey Keim. "Land Ownership and Women's Power in a Midwestern Farming Community." *Journal of Marriage and the Family* 39 (February 1979): 109-119.

(from 54 to 81%) the husband and wife were jointly concerned in decision-making. The author suggests that technical, economic, and social knowledge must be directed toward the whole family.

377. Abbott, Susan. " Full-Time Farmers and Week-End Wives: An Analysis of Altering Conjugal Roles." *Journal of Marriage and Family* 38 (February 1976): 165-174.

Correlates of decision-making in rural Kenyan domestic units are examined. Special attention is given to the contrast between ideal expectations and verbal reports of actual decision patterns. The author used Blood and Wolfe's (1960) testing technique for encouraging husbands and wives to report their relative power in family decision-making. The investigator saw the respondents twice, first to get their ideal response and, a second time, when she knew them better, to discover how things were actually done.

378. Barbic, Ana. "Farm Women, Work and Decision-Making: Yugoslav Experience." *Sociologia Ruralis* 28:4 (1988): 293-299.

Women participate in decision-making equally with men in Yugoslavia. However, the participation of rural women in decision-making bodies in agricultural co-operatives such as local communities, communes, republics, and the nation, where women have hardly ever been represented, is not satisfactory. Rural women, traditionally organized by agricultural co-operatives, will have to find their place within newly founded farm organizations.

379. Berlan Darque, Martine. "The Division of Labour and Decision-Making in Farming Couples: Power and Negotiation." *Sociologia Ruralis* 28:4 (1988): 271-292.

Questionnaires concerning division of labor and decision-making among farm couples in France indicated that big decisions dealing with domestic space, the upbringing of children, and farming are made in a fairly egalitarian manner. However, analysis of control in the sphere of domestic work and production shows that men, who are absent from the childraising sphere, have more control over it than women who are closely associated with it. The type of negotiation used by women aims at improving their position by way of accumulating control of the reproductive sphere and employing strategies to avoid domestic work.

380. Bokemeier, Janet, and Lorraine Garkovich. "Assessing the Influence of Farm Women's Self-Identity on Task Allocation and Decision Making." *Rural Sociology* 52:1 (1987): 13-36.

Data from this survey come from a statewide mail survey of 880 Kentucky farm women. Multivariate analysis examines the association of women's task performance and decision making with women's self-identity and other factors, such as farming background, type and size of operation, personal characteristics, and off-farm employment. The authors found that joint and egalitarian farm decisions do occur, but, overall, women indicate low levels of participation in farm decisions. Statistical information and a short list of references are included.

381. Burchinal, Lee G., and Ward W. Bauder. "Decision-Making and Role Patterns Among Iowa Farm and Nonfarm Families." *Journal of Marriage and the Family* 27 (November 1965): 525-530.

Examines the relative dominance of husbands' and wives' decision-making and family task performance among farm and nonfarm residents. The authors found that no consistent or marked differences appeared. Data from families living in Des Moines and in several small towns and rural areas in Iowa were used to answer these questions. Statistical information is included.

382. Colman, Gould, and Sarah Elbert. "Farming Families: The Farm Needs Everyone." *Research in Rural Sociology and Development* 1 (1984): 64-78.

Describes a long-term case study at Cornell University's Farm Family Project started in 1967 as an inquiry into how decisions are made in farm families. Intensive field interviews were conducted on family farms in New York State during 15 years, gathering farm and household information from 33 families. The project started out with 19 families, but the number of respondents expanded as families matured and children of the original families entered farming in various forms of family succession. Fifteen of the 19 original families have remained in farming; 7 of them have received successful intergenerational transfer.

It was found that the farmers who participated in the Cornell study make choices within the context of the household, and they are influenced by the household's needs and goals as well as by the resources available to them. The resources include land, water, credit, and family labor as well as information about agricultural practice. This information is sought first from neighboring farms and professional consultants.

The expansion of agricultural higher education to include women seems to have made a difference in farming families' decisions to include farm daughters and daughters-in-law in ownership and management of the farms. It was also found that agricultural choices are flexible and responsive to changes in the farm family decision-making environment.

383. Rogers, Susan. "Female Forms of Power and the Myth of Male Dominance: A Model of Female/Male Interaction in Peasant Society." *American Ethnologist* 2 (1975): 727-756.

Looks at how power is distributed between males and females in peasant societies. The author starts with the assumption that males are universally dominant, she then shows that this generalization is based on a body of definitions and models which deal with only a limited and male-oriented range of phenomena and are clearly contradicted by a body of empirical evidence of real power wielded by peasant women. Both men and women behave publicly as if males were dominant, but the fact is that peasant women actually wield a considerable amount of power. She then suggests a model to explain these apparent contradictions, in which male dominance is seen to operate as a myth, while a balance is actually maintained between the informal power of women and the overt power wielded by men. The author limits her discussions to contemporary peasant societies, especially European ones.

384. Rosenfeld, Rachel A. "U.S. Farm Women: Their Part in Farm Work and Decision Making." *Work and Occupations* 13:2 (May 1986): 179-202.

The data used in this study are from the 1980 Farm Women's Survey, a national survey of farm women designed and carried out by the National Opinion Research Center. The author examines the range of tasks and decisions in which the farm women take part and the women's own feelings about the places they hold on the farm. It was found that these women were involved in a wide range of farm tasks and decisions. The characteristics of the farm and the farm labor force affected the extent of their involvement as well as women's employment outside the home. The farm women tended to identify with their farm as a family unit and did not see themselves as a separate group called "farm women," and they considered themselves to be one of the farm's main operators.

385. Salamon, Sonya, and Ann Mackey Keim. "Land Ownership and Women's Power in a Midwestern Farming Community." *Journal of Marriage and the Family* 39 (February 1979):109-119.

Data on which this analysis is based are derived from field research on land tenure patterns in three rural communities in central Illinois. The ethnographic data presented demonstrate the organization of power within farm families and the manner in which this power is wielded. In all situations, the ability to use power in this community and behavior directed at its acquisition seem to depend directly or indirectly on the control of land. The factor of female land ownership and the issues which are related to the control of land are significant to the understanding of the organization of farm families. The author says "the significance of female control of land must be recognized by those who work with farm families and those who formulate public policies."

386. Sawer, Barbara J. "Predictors of the Farm Wife's Involvement in General Management and Adoption Decisions." *Rural Sociology* 38:4 (Winter 1973): 412-426.

Examined are predictor variables hypothesized to associate with the wife's involvement in decisions concerning the general management of the farm business and decisions leading to adoption of agricultural innovations. The author found that involvement in both type of decisions was related to the wife's information-seeking activities, her involvement in farm tasks, and other situational variables such as family size, income, and farm size. Data for the study were provided by 67 married couples living on farms in Lower Fraser Valley of British Columbia, Canada. The author suggests several considerations for designing educational programs for farm families.

387. Sharma, D. K., and Tej Ratan Singh. "Participation of Rural Women in Decision-Making Process Related to Farm Business." *Indian Journal of Extension Education* 6:1-2 (March-June 1970): 43-49.

Describes a survey of 170 farm women from 10 villages in India to determine the extent of participation of rural women in farm operations and decision-making related to farm business. It was found that women possessing small land holdings and belonging to the middle age group, having no formal education and coming from the lower caste, participated in farm operations in larger proportions than other women and that these women took part in the decision-making with regard to seed storage, winnowing, care of animals, and harvesting. However, the majority of women were never consulted in regard to the use of pesticides, application of fertilizers and manures, and use of weedicides.

388. Shortall, Sally. "Power Analysis and Farm Wives: An Empirical Study of the Power Relationships Affecting Women on Irish Farms." *Sociologia Ruralis* 32:4 (1992): 431-451.

Examines how the dispersion and structure of power in a rural situation are related to the invisible role occupied by farm women and investigates how rural structure and situations create a less powerful position for women and why few grievances arise. Data for this study were primarily collected by participant observation and interviews with twenty farm wives in a rural area in a midland county fifty miles south of Dublin. Questions were generally open-ended and the interviews were for the most part of an informal nature, each lasting from one-and-a-half to several hours. Women interviewed seemed to underestimate their farm involvement and overestimate husbands' household involvement and portray them as experts in all areas. It became obvious from this study that the male farmer benefits from the current situation; he holds the higher status position. However, it also seems that the exploited labor of the farm wife is essential for the survival of the family farm. The author points out how this is also emphasized by other researchers. The author asks "why is it women's labor that is undervalued?" It is clear that farm women's inferior position has benefits for the industry as a whole. If the contribution of farm wives were to be acknowledged and fair retribution made, it would put huge strains on the farming industry. Future research with larger samples is recommended.

389. Wilkening, Eugene A. "Change in Farm Technology as Related to Familism, Family Decision Making, and Family Integration." *American Sociological Review* 19:1 (1954): 29-37.

This study is mainly concerned with the manner in which variations in the family produce variations in certain decisions affecting farming. Evidence presented suggests that family relationships as indicated by integration, familism (the ascendence of family interests over the interests of the individual members), and father-centered decision-making have little direct influence upon the acceptance of innovations or improvements in farming. The data were obtained from 170 farm operators and their wives in south central Wisconsin. Several statistical tables are included.

390. Wilkening, Eugene A. "Joint Decision-Making in Farm Families as a Function of Status and Role." *American Sociological Review* 23:2 (April 1958):187-192.

The data was obtained by personal interviews of 614 farm operators and their wives, randomly chosen from six Wisconsin counties. The author wanted to prove two hypotheses: (1) joint involvement of husband and wife in major decisions is not associated with the social status of the wife as indicated by educational level and formal social participation; (2) the greater the degree of commercialization of the farm enterprise, the less joint involvement of husband and wife in major farm and home decisions.
The first hypothesis was not supported except under specific conditions, but there was more support for the second hypothesis that joint involvement of husband and wife declines with degree of commercialization.

391. Wilkening, Eugene A., and Denton E. Morrison. "A Comparison of Husband and Wife Responses Concerning Who Makes Farm and Home Decisions." *Marriage and Family Living* 25 (August 1963): 349-351.

One way of studying influences in decisions is to compare the reports of husbands and wives on specific types of decisions. The data for this study were gathered from 61 randomly chosen farm families of a central Wisconsin county. The distribution of responses by husbands as compared to those of wives show little differences in how specific types of decisions were made. The most striking fact was that thirteen out of the twenty-three decisions were ordinarily shared about equally by husband and wife. Major investments or commitments of money including buying land, borrowing money, buying a car, buying life insurance, and how much money to save tend to be the concern of both husband and wife.

392. Wikening, Eugene A., and Eugene Lupri. "Decision-Making in German and American Farm Families: A Cross-Cultural Comparison." *Sociologia Ruralis* 5:4 (1965): 366-385.

The present comparative analysis is based upon interviews of 505 farm families in Wisconsin and 227 interviews from peasant families in Hesse, West Germany. The authors discuss some major economic and social characteristics of the two study groups, family composition, and family contribution to farm enterprise. The study reveals that the Hessian women are more involved in farm chores and field work than the farm women in Wisconsin. Among the Wisconsin farm wives there is more involvement in decisions involving land and machinery acquisition but less in operational decisions pertaining to the farm. Several statistical tables and a list of references are included.

393. Wilkening, Eugene A., and Lakshmi K. Bharadwaj. "Dimensions of Aspirations, Work Roles and Decision-Making of Husbands and Wives in Wisconsin." *Journal of Marriage and the Family* 29 (November 1967): 703-711.

Presents the findings from a statewide study of 500 Wisconsin farm families selected by a multistage area-probability sample design by the Wisconsin Survey Research Laboratory. The data were gathered from both husbands and wives who were both native born and had similar national and religious backgrounds. The authors attempt to delineate the dimensions of aspirations, task allocation, and involvement in decisions, and it was found that the value placed on specific goals by one spouse is different from that placed by the other spouse. They also found that there is a specialization in decision-making as well as in the performance of tasks with joint involvement in certain areas. Many statistical tables are included as well as some references.

394. Wilkening, Eugene A., and J.C. van Es. "Aspirations and Attainments among German Farm Families." *Rural Sociology* 32:4 (1967): 446-455.

Attempts to explore the interrelationship between aspirations for the home and for the farm with attainments in the farm and home areas among a sample of German farmers. Data were gathered in six villages in the Landkreis Marburg, Hesse, Germany. The households included farms of 2.5 hectares or more. The findings indicate that there is an interplay among motivation, resources, and attainments as they involve farm and family institutions in which the resources and the attainment of one are contingent on the other.

395. Wilkening, Eugene A., and Lakshmi K. Bharadwaj. "Aspirations and Task Involvement as Related to Decision-Making among Farm Husbands and Wives." *Rural Sociology* 33:1 (1968): 30-45.

Measures of the dimensions of aspirations, allocation of tasks, and involvement in decisions of husbands and wives are developed with the use of factor analysis. Interrelations among these dimensions indicate that the involvement of husbands and wives in farm, home, and family decisions is influenced by their task involvement and aspirations. Aspirations of wife and husband in the home area tend to affect the wife's involvement in farm decisions, but other aspirations have little effect on decision-making patterns.

396. Wilkening Eugene A., and Sylvia Guerrero. "Consensus in Aspirations for Farm Improvement and Adoption of Farm Practices." *Rural Sociology* 34:2 (1969): 182-196.

The data were obtained from a statewide multistage probability sample of 505 Wisconsin farm families in 1962. It was found that the wife's aspirations are more highly associated with mechanization than those of the husband, which suggests that she views labor-saving equipment as improvement more than he does. The husband sees dairy management and soil conservation practices as constituting improvement more than the wife. The hypothesis that the adoption of improved farm practice is higher when both husband and wife have high aspirations for farm improvement than when only one or neither does was supported by this research.

WOMEN'S ROLE IN AGRICULTURAL POLICY IMPLEMENTATION

Several researchers feel that all the research projects on women in agriculture have helped a great deal in making women farmers visible on the international level. The Decade for Women has also helped put forward resolutions to direct more resources to women farmers--and Africa, where women produce more than 60% of the food, has been in the news and media.

Kathleen Staudt, one of the researchers interested specifically in the political issues related to women farmers, points out in one of her studies that much credit goes to the United Nations Economic Commission for Africa, where the Human Resource Development Division prepared "Women: The Neglected Human Resources for African Development" published in the first special issue on African Women in the *Canadian Journal of African Studies*. She argued that it is not recognized that women are responsible for more than 60% of the agricultural labor in Africa and that the women do not receive the training needed nor do they have access to the agricultural resources. Staudt points out that although information was widely disseminated and helped to start new research projects, twenty years passed before some action developed on the national and international level. Projects funded by international development are finally responding to the demands of more resources and training for women. Professor Staudt published an important textbook, *Managing Development: State, Society, and International Contexts,* stressing political context and process, that will help future researchers to understand the problems and see the need for action.

There are studies not only from Africa but from all parts of the world pointing out the hard work of women farmers and their unfair share of the rewards. Even in countries in Eastern Europe, where one would think equality existed, women do not often get the top jobs in management of the communal farms. Planners have to start from the beginning to include women in their projects if the outcome is to be beneficial, not only for the participants but for the countries as a whole.

397. Adekanye, Tomilayo O. "Women in Agriculture in Nigeria: Problems and Policies for Development." *Women's Studies International Forum* 7:6 (1984): 423-431.

Discusses the nature of women's involvement in agriculture in a case study undertaken in three selected areas of Nigeria in 1979. A sample of 200 women was studied from each of the three survey areas. A short review of the place of agriculture in the economy is given, and the role of women in Nigerian society is assessed. The major conclusion that emerges is that women are involved in all aspects of agriculture from production to processing and trade. An employment, educational, and income strategy needs to be devised for women in order to enhance their contributions to agricultural and rural development in Nigeria.

398. Agarwal, Bina. "Who Sows? Who Reaps? Women and Land Rights in India." *Journal of Peasant Studies* 15:4 (1988): 532-581.

A survey article tracing women's past and existing rights to land in law and in customary practice across communities and regions in India. Seeks to identify the factors impinging on women's ability to claim, control, and self-manage land today. Modern legislation has yet to establish full gender equality in law or to permeate practice. Customary practices governing marriage, residence and female functioning of official agencies, etc., variously obstruct women in claiming their legal share of functioning as independent farmers, although the nature and incidence of these factors differs cross regionally. Some statistical information and a long list of references are included.

399. The Asian and Pacific Development Center (APDC). *Agricultural Change, Rural Women and Organization: A Policy Dialog*, Proceedings of the APDC-ACWF International Seminar. Beijing, China, 29 October- 8 November 1986. Kuala Lumpur: Asian and Pacific Development Center, 1987.

This Policy Dialog is the result of two years of planning and research of the program "Agricultural Change, Rural Women and Organization," designed partly to document examples of the effect of agricultural change on rural women in India, Sri Lanka, Thailand, China, Malaysia, Indonesia, and the Philippines. The research has pointed to a number of areas where women's position has suffered due to the effects of agricultural change. The Policy Dialog confronts some of those areas and examines possible interventions by relevant government agencies which may offset and improve these consequences. The program of the Policy Dialog and a list of the participants are included.

400. Barlett, Peggy F. "Adaptive Strategies in Peasant Agricultural Production." *Annual Review of Anthropology* 9 (1980): 545-573.

Reviews methods of agricultural production strategies and how they help us understand production processes and change in a peasant society, which leads to the general study of human culture and societies. Through closer attention to the diversity within peasant communities, delineation of the relevant variables that determine different production strategies has become possible. Combining the local idiosyncrasies of time and place with national economies, political, and social institutions and forces permits the identification of the range of possible production choices. Future research is needed to explore the long-range impact of production strategies and adaptive processes, particularly in light of the ever-increasing damage to the world ecosystem and the rapid pace of proposed technological changes in agriculture. An extensive list of references is included.

401. Beoku-Betts, Josephine. "Agricultural Development in Sierra Leone: Implications for Rural Women in the Aftermath of the Women's Decade." *Africa Today* 37:1 (Winter 1990): 19-35.

One of the most critical issues addressed during the period of the Women's Decade concerned the effectiveness of national development policies and institutional machineries to facilitate the empowerment of women. The present economic situation in Sierra Leone, the need to take more progressive action to reduce poverty, food shortages, unemployment, and wage restrictions must become a priority for development. Women not only share these problems, but they perform critical roles in the processing, preservation, and marketing of food, and in the capacity of the family to survive under deteriorating economic conditions. Given their importance and their relative powerlessness, policy interventions to address these problems must therefore be more sensitive to these concerns.

402. Clark, Carolyn M. "Land and Food, Women and Power, in Nineteenth Century Kikuyu." *Africa* 50:4 (1980): 357-369.

Using data on the Kikuyu of the nineteenth century, the author shows that an understanding of the political economy which emphasizes the process by which productive activities are built into political relations avoids arbitrary distinctions between prestige and subsistence economies and women's and men's domains. She asks, "Were women controllers of resources or themselves resources controlled by men?" The right to make decisions, the spiritual and economic power to enjoin compliance, and the expectation that wives obey their husbands together set up a dynamic of sometimes conflicting expectations which gave rise to decisions that at times affirmed women's control over resources and at others, men's power over women. The Kikuyu women emerged as actors with control over resources vital in a system in which relations of production enter into political strategies and are built into the social relations of power.

403. Chen, Marty. "A Sectoral Approach to Promoting Women's Work: Lessons from India," *World Development* 17:7 (1989): 1007-1016.

Women are left out of sectoral economic planning, because government policy makers do not view women as productive workers. The author suggests that donor agencies could help bring women into the mainstream of the planning process and the economy by commissioning studies of women's work by sectors and by supporting sector-based pilot projects that involve women. This has been done successfully in India in two fields, dairying and silk production.

404. Davison, Jean. "Without Land We Are Nothing: The Effect of Land Tenure Policies and Practices upon Rural Women in Kenya." *Rural Africana* 27 (1987): 19-33.

In Kenya, women are the demographic majority in rural areas, produce more than 80% of the food crops, contribute substantially to the production of cash crops, but own only 5% of the land nationwide. Related to differential patterns of land use and ownership within the Kenyan peasantry are gender-based issues that have historically shaped women's direct or indirect access to land and, hence, their ability to feed their families.

National land development policies are of two sorts: those designed to increase economic growth and efficiency and those intended to increase equity. In Kenya policies have followed the growth model to women's disadvantage. Single women in particular have suffered. The solution to women's lack of access to land requires decisions at the national level that put into practice laws guaranteeing a woman's right to inherit land as a daughter.

Legislation is also needed to ensure that widows, who currently have no legal protection, receive the right to inherit their husband's property. Finally, policies must be advanced that make available to women, regardless of marital status, capital for the purchase of land.

405. Deere, Carmen Diana. "Rural Women and State Policy: The Latin American Agrarian Reform Experience." *World Development*. 13 (September 1985): 1073-1053.

Reviews 13 Latin American agrarian reforms and finds most have directly benefited only men. The author argues that this is mainly because of the common designation of "households" as the beneficiaries of an agrarian reform and the incorporation of only male household heads into the new agrarian reform structure. It is therefore necessary that women also be designated as beneficiaries. Women as well as men must be given access to land or the opportunity to participate within the agrarian cooperatives or state farms promoted by an agrarian reform.

406. Deere, Carmen Diana and Magdalena Leon, eds. *Rural Women and State Policy: Feminist Perspectives on Latin American Agricultural Development.* Boulder: Westview Press, 1987.

Central questions here are on whose terms and under what terms rural women will participate in the solutions to the Latin American economic crisis. Providing a synthesis of what is known about women's participation in Latin American agriculture, the authors present an overview of Latin American development and agricultural policies, highlighting the ways in which these policies affect rural women. The book is divided into two parts, covering models of development, agrarian policy, and the economic crisis and looking at comparative perspectives on development initiatives. Several references follow each chapter.

407. Dey, Jennie. "Development Planning in the Gambia. The Gap Between Planners' and Farmers' Perceptions, Expectations and Objectives." *World Development* 10:5 (1982): 377-396.

Challenges the prevalent view that irrigation development largely depends on engineering, agricultural, and managerial inputs and argues with reference to Gambia irrigation projects that new technologies may be adapted by farmers in ways that are incompatible with planners' objectives. It is suggested that planning should proceed in three steps. First, the planners should understand how the existing farming system operates as a whole and try to understand the interrelationships between technical factors and the supply of land and labor and the relative economic returns of the different crops. Second, they need to take great pains to find out what the farmers value in this system, and, third, instead of proposing changes to farmers and offering bribe-like incentives in the form of credits or free inputs, they should first ask the farmers how they see their problems and what changes the farmers would welcome.

408. Due, Jean M., and F. Magayane. "Changes Needed in Agricultural Policy for Female-Headed Farm Families in Tropical Africa." *Agricultural Economics* 4 (1990): 239-253.

Decline in per-capita agricultural production has been reversed in tropical Africa by the structural adjustment program which increased producer prices, liberalized marketing and devaluated currencies. The 30% of smallholder farmer households which are female-headed will not be assisted much due to their special constraints of shortage of labor and credit, lack of extension visits, and appropriate labor saving technologies. Documents the significant differences in resource endowments and suggests ways in which agricultural and extension policy could be changed to benefit this group. References are included.

409. Due, Jean M. "Agricultural Policy in Tropical Africa: Is a Turnaround Possible?" *Agricultural Economics* 1 (1986): 19-34.

Agricultural output in sub-Saharan Africa has increased at 2% yearly, but population growth has surpassed agricultural production, and production per capita is lower now than at independence. The author reviews the internal and external factors which have contributed to this depressed situation and highlights the major policy changes needed.

410. Ensminger, Jean. "Economic and Political Differentiation among Galole Orma Women." *Ethos* 52 (July 1987): 28-49.

Formal and informal positions of power held by Orma women are analyzed in relation to effectiveness of women's influence over production decisions. Access to wealth was not a sufficient condition for the improved economic and political state of women. It was concluded that if women are to have a greater role in resources management and political participation they must have direct access to productive capital, the market, and the political arena.

411. Geisler, Charles C., et al. "The Changing Structure of Female Agricultural Landownership, 1946 and 1978." *Rural Sociology* 50:1 (1985): 74-87.

Female, male, and joint male/female farmland ownership is compared for the United States in 1946 and 1978. Findings indicate that women's share of farmland, measured in terms of average acreage and percent of total farmland, has increased somewhat over 30 years despite the disproportionately large share retained by male and joint male/female owners in both time periods. Tenure and modes of acquisition of the farmland differ with gender, probably due to the late age at which most women acquire land. The need for expanded research on women as owners of farmland is clarified.

412. Haney, Wava G., and Lorna Clancy Miller. "U.S. Farm Women, Politics and Policy." *Journal of Rural Studies* 7:1-2 (1991): 115-121.

Explores the emergence, orientation, and initial influence of the Women Involvement in Farm Economics (WIFE) organization, compared to that of their foremothers, who were members of general farm organizations and associated political movements. The authors found that while the level of farm women's participation in farm politics and policymaking has changed, like their foremothers, they are oriented toward family farm and farm family welfare, education, and unity issues.

413. Kabeer, Naila. *Reversed Realities: Gender Hierarchies in Development Thought*. London: Verso, 1994.

Provides a reassessment of development theory and policy, beginning with an exploration of different notions of development. Traces then the emergence of women as specific category in development thought and examines alternative frameworks for analyzing gender hierarchies, such as Triple Roles Framework(TRF), which seeks to establish a bridge between women's needs and planning process. Identifies the household as the primary site for power relations and compares different approaches to the concept of the family unit. The author urges women to be agents of their own change. An extensive bibliography is included.

414. Kandiyot, Deniz. "Women and Rural Development Policies: The Changing Agenda." *Development and Change* 2:1 (January 1990): 5-22.

Reviews and evaluates policies directed at rural women in the Third World as reflected in the Women in Development (WID) research and policy documents. An attempt is made to show how the agenda of mainstream WID research and policy formulation has closely followed, reflected, and responded to changing international priorities in matters of development assistance in a manner that leaves crucial redistributive and political issues hardly addressed and unresolved.

415. Kardam, Nuket. *Bringing Women in: Women's Issues in International Development Programs.* Boulder, CO: Lynne Rienner Publishers, 1991.

The women's movement, together with the environmental movement, is changing the way development issues are defined. In the 1970s, an international social movement emerged that had as one of its objectives the implementation of more gender-sensitive policies. This social movement has documented the adverse impact of agency policies on poor women of developing countries, and the problems were identified as ingrained attitudes, values, and perception among development personnel as well as a lack of information and data on women and lack of resources allocated to women. To correct this situation three major policy recommendations were made: (1) resocialization of personnel through training and education programs, (2) redirection of research and data collection to include women, and (3) allocation of resources to employ more female development professionals. The author takes a closer look at women in development (WID) activities and feels that the international women's movement as a social movement has been able to penetrate the development assistance regime, but is restrained from turning general ideas into specific policies.

416. Keller, Bonnie, and Dorcas Chilila Mbewe, "Policy and Planning for the Empowerment of Zambia's Women Farmers." *Canadian Journal of Development Studies* 12:1 (1991): 75-88.

Argues that the bureaucracy of the Ministry of Agriculture in Zambia can be gender sensitized to increase women farmers' access to resources. One important area is research which provides quantitative data on women's contribution to different types of agricultural production. It is vital to have such information in order to demonstrate to agricultural policy makers and planners that women make specific contributions and that they should be a target group to whom supportive services are directed. The women's level of political participation is low, and empowerment is not high on women's agenda. Hence, a supportive bureaucratic structure is necessary to lay the basis for women's future political participation to challenge their subordination.

417. Lofchie, Michael F., and Stephen K. Commins. "Food Deficits and Agricultural Policies in Tropical Africa." *The Journal of Modern African Studies* 20:1 (1982): 1-25.

Examines four approaches to Africa's food deficits and suggests a series of critically important reforms which can best be understood as broad guidelines for future agricultural development in Africa. Genuine empowerment of the peasantry is recommended in order to enable rural producers to affect the political process, the character of the marketplace, and the administration and implementation of foreign aid. Implementation of land tenure policies is recommended to preserve the positive features of small-scale peasant agriculture and provide free access to commodity markets through by-passing the large firms which presently dominate the international marketplace in Africa's major export products. Introduction of agricultural policies that are more fully influenced by sensitivity to environmental constraints is also recommended. The authors feel that the planning of African agriculture has separated environmental from development issues and that the result has been agriculture policies which devastate the land and drastically diminish its value as an economic resource for future generations.

418. Moock, Peter R. "The Efficiency of Women as Farm Managers: Kenya." *American Journal of Agricultural Economics* 58:5 (December 1976):831-835.

Investigation of possible differences between male and female farm managers in the possession of and means of acquiring technical information relevant to agricultural production of maize, the staple commodity of Kenya. While the impact of schooling on output is greater for the women than for the men, other factors remain the same. Another striking finding has to do with the impact of the extension services on farming. The women seem not to benefit, as the men do, from extension contact, perhaps due to the marked male orientation of the services provided by Kenya's Ministry of Agriculture.

419. Morvaridi, Behrooz. "Gender Relations in Agriculture: Women in Turkey." *Economic Development and Cultural Change* 40:3 (April 1990): 567-586.

Examines the impact of technological change on rural women in Turkey. Looks at state-supported technical change through various changes of government policy, from inward growth to structural adjustment and liberalization, and the adoption behavior of the family farms. The transition from subsistence farming to commercial production alerts the values and norms not only of production relations but also those of everyday life. The author found that government policies did not consider women's position in agriculture.

420. Moser, Caroline O. N. *Gender Planning and Development: Theory, Practice and Training.* London: Routledge, 1993.

Describes the development of gender planning as a legitimate planning tradition in its own right. It is of critical importance to inform policy through the formulation of gender policy at international and national levels as well as its integration with sectoral planning. In addition more appropriate, that is, gender-aware, planning procedures must be developed. Severe problems have been experienced with gender issues, especially in ensuring that policies were implemented. The purpose of this book is, therefore, to assist in understanding more comprehensively gender policy and planning process and the constraints on the implementation of practice. An extended bibliography and name and subject indexes are included.

421. Newbury, Catherine M. "Ebutumwa Bw'Emiogo: The Tyranny of Cassava. A Women's Tax Revolt in Eastern Zaire." *Canadian Journal of African Studies* (1984): 35-54.

Explores the conditions which fostered the movement of more than 100 women to converge on the administrative office of Buloho Collectivity in Zaire in opposition of taxes being levied on the cassava and peanuts they transported to sell at the market. Includes also analysis of the circumstances which may encourage collective solidarity among women in Africa, enabling them to engage in political confrontation with men, and the constraints to such action in the political context of Zaire's Second Republic.

422. Orvis, Stephen. *Men and Women in Household Economy: Evidence from Kisii*. Nairobi, Kenya: Institute for Development Studies, 1985.

Analyzes household economy to understand the potential for agricultural development policy, particularly extension policy aimed at smallholder farmers in Kenya. A proper understanding of that economy must start with an understanding of the different structural positions and access to resources of men and women. Particular attention is paid to the Ministry of Agriculture's attempts at aimed at improving the effectiveness of agricultural extension work.

423. Pausewang, Siegfried. *Who Is the Peasant?: Experience with Rural Development in Zambia*. DERAP Working Paper. Bergen, Norway: Chr. Michelsen Institute, 1987.

A new policy announced by President Kaunda will restructure the national economy to produce what the population consumes and ask people to consume only what they can produce. It relies on agriculture to increase production through the efforts of the majority of small farmers. The author argues that the new economic recovery program should primarily support rural women to create growth where it benefits the majority of the rural poor.

424. Pausewang, Siegfried, et al., eds. *Ethiopia: Rural Development Options*. London: Zed Books, Ltd. 1990.

Since the land reform of 1975 the Ethiopian government has embarked on a consistent policy of socialist transformation of agriculture. It formed peasant associations and service cooperatives, created a new marketing system, state farms, cooperative farms and launched a major resettlement program. Despite 15 years of socialist development, Ethiopia is far from achieving food self-sufficiency, and the agrarian crisis in Ethiopia has prompted major international debates on the causes of the disaster. In 1988 at the International Conference of Ethiopian Studies in Paris, some 20 contributors volunteered to write one chapter each on their respective field research in rural Ethiopia, and a group of editors put together the chapters that make up this book. Much time and discussion have been devoted to a discussion of rural society and agricultural policy as they were before the Central Committee introduced the recent changes; however, the editors believe the new developments have made this book all the more topical. In chapter five Hanna Kebede summarizes women's position and talks about property ownership, decision making in the household, and women's political and economic participation. Helen Pankhurst analyses the question of continuity and change in the world of women from Menz, northern Shoa in chapter 13. A bibliography, glossary, and index are included.

425. Randolph, Sheron, and Rickie Sanders. "Constraints to Agricultural Production in Africa: A Survey of Female Farmers in Ruhengeri Prefecture of Rwanda." *Studies in Comparative International Development* 23:3 (Fall 1988): 78-98.

Findings from a survey of 192 women farmers in Ruhengeri Prefecture of Rwanda in east central Africa suggest that Rwanda is experiencing serious demographic and environmental problems and that the traditional mechanism of adaptation can no longer be relied upon to bring about equilibrium. The women of Rwanda are the conduits of change since they contribute to the largest percentage of agricultural labor and have the most responsibilities for operation of farms and production of agricultural output. The author suggests that special attention must be focused on policy initiative geared toward reducing population growth, restructuring extension services, and reducing gender bias.

426. Rogers, Barbara. *The Domestication of Women: Discrimination in Developing Societies.* New York: St. Martin's Press, 1980.

This study is about planners trained in the Western tradition and the impact of their ideas about women in general on poor women in the Third World. The author says "she uses the word planner in a very broad sense to include all those who determine the formulation, design and execution of development policies, programs and projects." The book is divided into three parts and treats first problems of perception, then discrimination in development planning, and, in the third part, looks at the effect of development planning on women and their dependents. In the introduction the author laments the lack of books with reference to women and their work, the division of labor, and gender roles. Very few references are included.

427. Safilios-Rothschild, Constantina. "The Persistence of Women's Invisibility in Agriculture: Theoretical and Policy Lessons from Lesotho and Sierra Leone." *Economic Development and Cultural Change* 33:2 (1985): 299-318.

Examines "the institutional, normative, and attitudinal biases in Lesotho and Sierra Leone that help perpetuate and reinforce women's invisibility in agriculture and therefore perpetuate the image of agriculture as a male domain requiring agricultural development activities to be oriented to men." The author concludes that women remain invisible because the magnitude of rural women's work is not documented.

428. Staudt, Kathleen. "Agricultural Productivity Gaps: A Case Study of Male Preference in Government Policy Implementation." *Development and Change* 9 (1978): 439-457.

Examines the differential impact of policies on the sexes and draws implications about differentiation for the productivity of sex groups. The failure to recognize women's interests in societies with separate spouse economic activities or separate spouse residences as a result of migration significantly conceals our understanding of the government distributive process. Persistent privilege of one group at the expense of another can result in real differences in economic productivity between those groups with important consequences for their political power and ultimate life chances.

Argues that if present trends continue, women's productivity as compared to that of men will lessen, with several detrimental consequences to the development process. A continued policy of extending services by virtue of a recipient's maleness rather than by merit or demonstrated innovativeness, suggests rather ominous implications for government ability to significantly maximize agricultural goals.

429. Staudt, Kathleen. "Bureaucratic Resistance to Women's Programs: The Case of Women in Development" in *Women, Power and Policy*, Bonaparth, Ellen, ed. New York: Pergamon Press, 1982, pp.263-281.

Although bureaucratic structures are in place to provide for equal wages and equal opportunities, women's economic position relative to men's has eroded. Women's policy studies have alluded to implementation problems, but none has systematically examined the implementation process once policy pronouncements are in place. Monitoring is difficult and the data produced are not always reliable. Despite resource shortages, Women in Development (WID) has been able to formulate alliances and generate academic literature, which, in turn, sparks interest and builds credibility in the agency and in other institutions.

430. Staudt, Kathleen. "Class and Sex in the Politics of Women Farmers." *The Journal of Politics* 41 (1979): 492-512.

Data from this case study indicate that a good deal of differentiation among women has resulted from economic and political change. Patterns of agricultural service delivery have differential effects on the sexes, particularly among non-elite farmers. Two sample surveys are presented, one a cross-section of farmers (non-elite group) and another smaller sample from the elite class in Kenya. The results show that the most neglected clientele of the agricultural administration, non-elite women, bonded together to compensate for the discrimination. Elite women farmers experience equity in agriculture service delivery, and that equity reinforces both differentiation among women and the privileged position of the emerging elite class.

431. Staudt, Kathleen, and Jane S. Jaquette, eds. *Women in Developing Countries: A Policy Focus*. New York: The Haworth Press, 1983.

The essays in this book have also been published in the journal *Women and Politics*, v.2 (4), Winter 1982. "We are at a crossroad, facing key decisions about research and action," maintain the editors in the introduction. "Do analysts focus on the concrete strategies of putting lessons learned into practice of existing structure? Do such strategies involve complicity with an evolving structure?" are the questions the editors pose, and they have selected input from several perspectives. Tracy Ehlers analyzes the movement of independent female artisans working in a machine knitting process largely outside their control. Susan Rogers reviews macro-level studies from Tanzania, where feminist awareness among impoverished rural women is demonstrated. Three studies demonstrate "how we must understand the policy and program implementation institutions within which any transition toward more women sensitive change is to occur." Hanna Papanek in her chapter argues that "all policy issues are women's issues, but that policy must be sensitive to differentiation among women." The last chapter includes a selected bibliography on Women, Development, and Public Policy compiled by Sandra Danforth (See entry 622).

432. Staudt, Kathleen. *Women, Foreign Assistance, and Advocacy Administration.* New York: Praeger Scientific, 1985.

The author is looking at the development of administration or evaluation literature and is struck with the lack of attention to gender in the implementation or the distributive process. As a researcher moving upward analytically in the field from field agent to researcher in a large development agency, the author feels she has an understanding of the reasons for this curious inability to deal with gender in bureaucracy. The methodology of the study combines both quantitative and qualitative techniques of ethnographic field research. An extended bibliography is included.

433. Staudt, Kathleen. "Women Farmers in Africa: Research and Institutional Action, 1972-1987." *Canadian Journal of African Studies* 22:3 (1988):567-582.

The author agrees with other researchers that agricultural policy should be designed as if women are the major agricultural producers. The changing discourse on and wider audiences for research on women farmers suggests some promise of putting all the grand policy statements into action in national and international bodies. Women farmers' voices must be heard and their power exerted in bureaucratic, state, and political processes. Only then can women help create and determine a development process that is beneficial to them. An extensive list of references is included.

434. Staudt, Kathleen. *Managing Development: State, Society, and International Contexts.* Newbury Park, CA: Sage Publications, Inc, 1991.

This textbook stresses political context and process throughout, but its focus is bureaucratic politics and political relationships between people, their organizations, and the state in development programs and projects. Contains development cases, hypothetical situations, examples, and role-playing exercises (which are designed to build analytical skills) and group discussions based on holistic diagnosis of problems from different vantage points, from the so-called grassroots to the state and international economy, on development options and their management in organizations. The text explains and evaluates techniques in development management and stresses the diagnosis of political contexts and organizational politics more than techniques. References are listed after each chapter.

435. Stuart, Robert C. "Women in Soviet Rural Management." *Slavic Review* 38:4 (1979): 603-613.

Focuses upon women in top management in the Soviet rural sector. Notes the qualifications of those who hold these top positions, examines the paths by which they achieved power, and then compares their characteristics with those of all the women available for employment. The investigation was limited to what data could be found. It shows that women have been most prevalent in the lowest and least important managerial positions. To some degree this pattern has changed in the last twenty years, although one fact still stands out: women do not attain top positions, representing, at best, only 1 to 3% of the top managers. The role of women in agricultural management is even more limited than their role in industrial management.

436. Sudha Rani, P., et al. "Wage Differentials and Factors Governing Employment of Women in Agriculture." *Agricultural Situation in India* (July 1990): 249-252.

Examines the wage differentials, labor force, labor participation and share of females in total wages and total output and identifies the factors governing employment of women in paddy and cotton based cropping systems in India. It was found that there are differences in wages in contract method of employment, and it is suggested that Equal Remuneration Act should be strictly implemented for similar work performed by males and females. Several tables of statistics are included.

THE EDUCATION OF WOMEN IN AGRICULTURE

One of the most important facts about agricultural education and women is that, for the most part, the survival and health of people worldwide rest strongly upon the agricultural production of women farmers, who produce an overwhelming majority of the subsistence food supply. Yet, international investment in the agricultural education of women awaits rescue from a multitude of complex gender, political, legal, social, economic, and/or societal conditions to the status of urgency of highest priority.

The issue of women's agricultural education has a long history and reflects the extent to which policymakers went to insure that women had little or no access in this vital area. Only in recent years have policymakers taken steps to amend legislation and societal traditions for full human resource development. As a result, scholars have begun addressing a wide spectrum of concerns related to the female aspects of farming, food supply, agricultural education, and economic development.

Coming to the forefront along with the development of women's movements, the education of women in agriculture as a research topic dates mainly from the 1970s in a few countries. The extent of literature which specifically addressed the education of female farmers is evidenced in the scope of research and has not yet become a central focus of research. Even in the 1990s, there is much more information about the lack of education than about the plans, implementation initiatives, achievements of agricultural education programs for women or the substantial inclusion of women in agricultural education programs.

Tuition fees, societal traditions of educating only males, legal restrictions, inadequate family support, and lack of access to higher education are among the factors that restrict/limit agricultural education of women.

Without adequate access, education, and training in agriculture for the rural populations as a whole, nations cannot attain full fruition of human resource development in the production of food supply, improve the health of the nation, and alleviate poverty. Women provide over 60% of agricultural labor in many nations, yet the importance of education and technological training was not addressed until recent decades. Although great strides have been made in some regions since the 1960s, most are far from full inclusion of women in the formal educational structure.

The agricultural education of women falls roughly into three areas: formal, informal, practical family tradition. Most women farmers learn from the oral practical family traditions as a significant percent of them are illiterate.

Extension services provide some of the most consistent and continuous formal and informal agricultural education for females. Such services are innovative and diverse to accommodate traditions and customs in different regions ranging from onsite training, demonstrations, workshops, and programs to distance education via radio, video, visiting scholars, and correspondence courses.

With the establishment of laws in the United States that opened access for women to formal and informal programs, academic institutions expanded curricula and student recruitment to include females who had been excluded from postsecondary education. The 4-H Clubs and Future Farmers of America (FFA), and Farmers' Clubs provided activities that helped to motivate young women to consider agricultural education as a profession. These and similar organizations in the international community have provided informal training and encouragement to girls and young women. Such activities along with oral traditions in farm families have been positive indicators in gender roles for agriculture.

By the 1970s, the topic of women in agriculture had captured the attention of media to the extent that the *Agricultural Education Magazine* devoted the June 1975 issue to women in agriculture, and the March-April 1981 *Africa Report* was devoted to women in Africa. Several articles in each issue addressed the education and training of women and included a strong emphasis on better understanding, expanded research, involvement of women in policy development, and greater access and support for women as significant contributors to agricultural and economic productivity as well as better health for members of the global community.

The selected annotated bibliography which follows is a unique sampling of relevant works that reflect critical scholarship as well as popular subject related materials, with emphasis on sources that stimulate further research, greater support for the education of women farmers, and a global unity of purpose to recognize women for the significant societal contributions they make as major agricultural producers worldwide.

437. *Agricultural Extension: The Next Step.* Agricultural and Rural Development Series. Washington, D.C.: World Bank, 1990.

Emphasizes extension service as an effective means for providing education to a large number of small-scale farmers and agriculture as a solution to poverty. Sees extension divisions as important providers of information, rural advisory facility, and a vital information network. Shows the World Bank to be a major contributor to extension development in Third World countries. Sets out principles, proposes functional mechanisms and management practices for translating the principles into action.

438. African Training and Research Center for Women (ATRCW) Staff. "African Training and Research Centre for Women: Its Work and Program." *Africa Report* 26:2 (March-April 1981): 17-21.

Reports that rural women's access to land, training, improved technology, and off-farm income-generating activities is being adversely affected by mechanization, commercialization, and wrongly conceived development programs. Women continue to be ignored in national development planning in many countries. Identifies the legacy of colonialization as one of the main causes of the exclusion for African women from dynamic areas of social, economic, and educational change by promoting inappropriate models whereby women are relegated to "feminine" occupations and economically dependent on men. Describes activities of ATRCW, long-term objectives of the Center, research, training programs, establishment of a political network for advancement of women, and international seminars.

439. Bagchee, Aruna. *Agricultural Extension in Africa*. World Bank Discussion Papers No. 231, Africa Technical Department Series. Washington, D.C.: World Bank, 1994.

Provides useful information for development planners worldwide to encourage sustained agricultural growth. Summarizes deliberations and suggests policy implications from the January 1993 workshop for Anglophone countries and Francophone countries. Encourages specific education and support programs that reach women farmers, and highlights samples of such efforts to reach them via Farmer Clubs in Zambia and Malawi.

440. Baksh, Michael, et al. "The Influence of Reproductive Status on Rural Kenyan Women's Time Use." *Social Science and Medicine* 39:3 (1994):345-354.

Argues that agricultural development programs in sub-Saharan Africa must include a broad-based understanding of the cultural and social realities of women's lives. Recommends the development of closer links between agricultural programs and family planning efforts to reduce overall workload and allow greater flexibility in responding to their own family's economic and health needs. Points to the urgent need to provide women with current education about labor saving, high-yielding technologies that insure greater control over their time use and better response to farming crisis and opportunities.

441. Ballweg, J. A., and L. Li. "Employment Migration Among Graduates of Southern Land-Grant Universities." *Southern Rural Sociology Journal* 9:1 (1992): 91-102.

Studies the geographical mobility of 2,028 graduates of 15 southern land-grant universities. Compares those who accepted positions outside the state where they graduated (migrants) and those who remained within the state (non-migrants). Uses panel data from a survey conducted while students were enrolled in agriculture curricula at land-grant schools and a follow-up survey conducted a decade later. Although migrant graduates attributed more importance to work characteristics than economic reasons for accepting out-of-state-jobs, higher starting salaries and better benefits are found to be important factors associated with their move to another state. Differences were also detected for male and female graduates. Discusses theoretical interpretation and policy implications.

442. Bass, Herman M. "Women Agriculture Teachers." *Agricultural Education Magazine* 49:12 (June 1977): 281, 286.

Reports the results of two studies: (1) to determine attitudes of agricultural educators in the Commonwealth of Pennsylvania toward women as agriculture teachers, and (2) a California study of 20 women agriculture teachers. The Pennsylvania study found that women could perform well in all areas except large animal husbandry (50%) and agricultural machinery (25.7%), that women should not be limited to ornamental horticulture, that women had good classroom control, that women could manage shop courses, that women agriculture teachers could be accepted in the community, that most male agriculture teachers, supervisors, and teacher educators have a positive attitude toward females as agriculture teachers, and that women should be encouraged to enter the professional field of teaching vocational agriculture.

443. Bell, Lloyd C., and Susan M. Fritz. "Deterrents to Female Enrollment in Secondary Agricultural Education Programs in Nebraska." *Journal of Agricultural Education* 33:4 (Winter 1992): 39-47.

   Identifies critical considerations of female students which influenced their decision to enroll in agricultural education classes. Compares and contrasts parents', counselors' and agricultural education instructors' attitude toward enrollment of females in secondary agricultural education classes. Makes numerous recommendations, including formal and informal investigation to clarify critical career information and its delivery by students considering both traditional and nontraditional career opportunities; state leadership to provide inservice education to secondary educators on the theory and establishment of student support networks to facilitate nontraditional change; and identification of psychological considerations impacting individuals confronted with culturally nontraditional choices.

444. Bell, J. H., and U.S. Pandey. "The Exclusion of Women from Australian Post-Secondary Agricultural Education and Training 1880-1969." *Australian Journal of Politics and History* 36:2 (1990): 205-216.

   Examines aspects of the exclusion of Australian women from general and post-secondary education in agricultural and technical colleges between 1880 and 1969. Elaborates on the positive impact of the women's movement, which opened access to these institutions and resulted in increased enrollment of women by the end of the 1960s.

445. Cano, Jamie. "Male Vocational Agriculture Teachers' Attitude and Perception Toward Female Teachers of Agriculture." *Journal of Agricultural Education* 31:3 (Fall 1990): 19-23.

   A descriptive study of male vocational agriculture teachers' attitude and perception toward female vocational agriculture teachers in Ohio. Focused on the areas of job competency, leadership qualities, sexual discrimination sexual bias, and sexual harassment. Found that perceptions of sexual discrimination, sexual bias, and sexual harassment were evident from teachers, students, parents, the agricultural community, and employers; a very low percentage of male teachers have actually voted for female teachers for leadership positions within the professional organization. Female teachers are viewed as being able to fulfill leadership roles. Recommended that inservice education in the area of sex equity should be a priority of the state department of education staff, and that teacher education units should incorporate sex equity education into the pre-service program.

446. Carew, Joy Gleason. "A Note on Women and Agricultural Technology in the Third World." *Labour and Society* 6:3 (July/September 1981): 279-285.

Identifies problems associated with illiteracy which challenge farmers trying to achieve maximum agricultural productivity in many developing countries, some of which have suffered through generations of colonial underdevelopment. Women who supply a vast majority of agricultural export products are often overlooked when educational innovations and initiatives are advanced. Suggests that a restructuring of agrarian societies to recognize the valuable role of women will substantially increase productivity of the agricultural sector, increase health of the rural population, expand educational reforms, and enrich the nations.

447. Cebotarev, E. "A Non-Oppressive Framework for Adult Education Programmes for Rural Women in Latin America." *Convergence, An International Journal of Adult Education* XIII:1-2 (1980): 34-49.

Explores the relationship between education programs and social change and examines women's agricultural extension programs in 12 Latin American countries. Shows that most extended their curricula while avoiding consideration of new roles for women. Recommends a participatory mode of communication and research for education of women farmers and the improvement of domestic technologies as an important component of the education program.

448. Chakrapani, C. *Changing Status and Role of Women in Indian Society*. New Delhi: M.D. Publications, 1994.

Discusses education as one of the main factors improving in the status of women in India. Identifies formerly all male professions such as engineering in which women have made substantial gains at the higher education level. Urges greater support overcoming discrimination against women in rural areas.

449. Chaney, Elsa M. "Scenarios of Hunger in the Caribbean: Migration, Decline of Smallholder Agriculture, and the Feminization of Farming." *International Studies Notes* 14: 3 (Fall 1989): 67-71.

Reveals the unique educational status of women in Jamaica with its impressive record of attention to the nutritional quality of its population. Suggests that the high level of awareness of the causes and consequences of malnutrition provides a strong foundation for policy development through a linkage of food consumption and nutritional issues to agricultural policies. Recommends that policymakers move beyond planning which suggests that export products take precedence over subsistence indigenous agriculture to achieve food security and self-sufficiency concurrently. Encourages policymakers to take into consideration the specialized educational needs of women who cultivate most of the food used by the population.

450. Cooper, Barbara E, and Janet L. Henderson. "Career Perceptions of Women Faculty in Colleges of Agriculture." *NACTA Journal* (March 1989): 13-16.

Provides descriptive data on career perceptions for a sample of 218 women faculty in colleges of agriculture at U.S. land grant universities. Results of the seventy-two percent respondents along with interviews with fifteen of the women showed that positive experiences and encouragement from parents and teachers are important in career planning. Most of the women were satisfied with their career choices. Most of the women coordinated their professional time among teaching, research, and service responsibilities and balanced their personal time between family and individual needs. Includes references.

451. Cooper, Barbara E., and Janet L. Henderson. "A Profile of Women Scientists in Colleges of Agriculture." *NACTA Journal* (March 1988): 10-13.

A national study on women agricultural scientists in academic settings, which characterizes women scientists in colleges of agriculture at the 70 land grant universities in the United States. Found that women may be underrepresented on college of agriculture faculties, but they represent a young, dynamic, and successful group of scientists on those faculties. They have doctorates, receive tenure in six years, earn thirty to forty thousand dollars a year, teach and/or conduct research, publish research results, and have conducted an average of three funded research projects. When compared to U.S. Department of Agriculture statistics, these data indicate that there are more employment opportunities for agricultural scientists and engineers than there are graduates available for those jobs. Suggests that recruitment and retention of young women into science and agriculture must become a high priority.

452. Creevey, Lucy E. "Supporting Small-Scale Enterprises for Women Farmers in the Sahel." *Journal of International Development* 3:4 (1991): 355-386.

Explores reasons for success and failure of small-scale enterprise programs for rural women in the Sahel that focus on agriculture, animal husbandry, and manufacturing. Places strong emphasis on training village women in health, sanitation, preschool education. Discusses training and village-level agricultural research for women, starting agricultural research projects, and establishing other income-generating activities for women. Discusses training programs and workshops sponsored by the women's technical team.
Confirms that women's training made them conscious of balancing the family diet, better garbage management, and improved cleanliness, proper skills development, strengthened confidence; leadership skills through work on the technical training teams; and motivated women to assume a greater role in planning and ongoing training and retraining through seminars and workshops.

453. Curry, Charles. "Vocational Agriculture Programs--Emphasis on Female Interests." *Agricultural Education Magazine* 47:12 (June 1975): 270-271.

Discusses attitudes toward women students and the quality of instruction that they receive. Cites changes in societal expectations for women, including the women's movement, new legislation which provided access, and the implementation of career education as primary reasons for women's entry into agricultural education. Cites assumptions and unresolved problems of inequity.

454. Davison, Jean (with the women of Mutira). *Voices from Mutira: Lives of Rural Kikuyu Women.* Colorado: Lynne Rienner Publishers, 1989.

Examines the lives of seven rural women of different ages in Kenya as they mediate change in their individual and collective lives. Provides information about informal and formal training and impact of education on their development.

455. Dejene, Alemneh. "The Training and Visit Agricultural Extension in Rainfed Agriculture: Lessons from Ethiopia." *World Development* 17:10 (1989):1647-1659.

Provides an analytical framework for monitoring and evaluating extension projects through an examination of the applicability of training and extension visits in Ethiopia. Reveals training and visit system's effectiveness in disseminating innovations and increased production among contact farmers, upgrading extension agents' skill, and imparting valuable lessons for other extension systems in Ethiopia. Makes recommendations for improvements in applications by bringing women into all levels of training and visit system's structure from which women are noticeably absent. Supports the continuation of the system in fertile regions of Ethiopia but not in resource-poor and drought-prone regions of the country, which still pose a major unresolved problem for extension service agents.

456. Dhanakumar, V. C., and J. Lin Compton. "A Comparative Analysis of Institution Building for Agricultural and Rural Development in the Developing Countries: A Case Study." *Quarterly Journal of International Agriculture* 33:2 (1994): 192-204.

Focuses on development of better understanding about the Krishi Vigyan Kendra (KVK-Farmer Science Centers) in India as innovative institutions for vocational training of farmers/villagers and field level extension workers. Discusses current forces that impact the relevance and effectiveness of the centers as institutions for adult and extension education. One hundred five of the proposed 443 centers have been established. Based on an exploratory, descriptive, and analytical study that utilizes qualitative and quantitative data.

457. Dillingham, John M., et al. "Perceptions of Texas Agriscience and Technology Teachers Regarding Influence of Gender in Nontraditional Agricultural Mechanics Programs." *Journal of Agricultural Education* 31:1 (Spring 1993): 33-39.

Determines whether gender affects secondary and postsecondary teacher and student participation in agricultural mechanics programs, preferences of agriscience teachers, program descriptors which may represent limitations to gender equity in agricultural mechanics programs. Concludes that traditional teacher preferences for instructional areas indicate that equity has not been achieved in agricultural mechanics, that college courses have greater impact on the general mechanical knowledge of females when compared to males, and that males receive greater level of benefit from industry work experience and from full-time or part-time teaching experience. Recommends that female teachers who excel in and are interested in teaching agricultural mechanics be identified and publicized as role models for students and other teachers. Activities which enhance general mechanical knowledge should be made more readily available to both sexes, and curricula should be evaluated to determine and recommend improvements. Instructional strategies which encourage equitable enrollment should be developed and implemented; and increased emphasis on equitable access and participation should be implemented in high school and teacher education programs and workshops.

458. Doebbert, Jan. "Management Instruction for Farm Women-Learning From Experience." *Agricultural Education Magazine* 67:3 (September 1994):18-19, 22.

Reviews an experiential method of teaching management for farm women enrolled in the Alexandria Technical College (Minnesota) "Women in Agriculture" program. Based on L. Joplin's characteristics of experiential learning: (1) student-based rather than teacher-based, (2) personal, not impersonal, nature, (3) process and product orientation, (4) evaluation for internal and external reasons, 5) holistic understanding and component analysis, (6) organized around experience, (7) perception-based rather than theory-based, and 8) individual-based rather than group-based.

459. Drum, Sue, and H. Ellen Whiteley. *Women in Veterinary Medicine: Profiles of Success.* Ames: Iowa State University Press, 1991.

Profiles experiences of twenty women in a profession once considered a male domain--veterinary medicine. Chronicles the history of women's entry into the profession; indicate that women's salaries are averaging $1000 less than men. By the 1980s women were beginning to assume leadership roles in the American Veterinary Medical Association and the American Animal Hospital Association, and there was a prospect for greater authority in the veterinary hierarchy in the next twenty years.

460. *Enhancing Women's Participation in Economic Development*, A World Bank Policy Paper. Washington, D.C.: World Bank, 1994.

Points to steps that can be taken to correct inequities caused by barriers of low investment in female education and health. Discusses women's role in achieving social justice, in increasing agricultural production, in improving natural resource management for economic growth, and in reducing poverty. Suggests a broader approach toward a gender and a development strategy that takes into account the relative roles and responsibilities of women and men in a manner that produces long-term change in actions and attitudes that improve the conditions of women.

461. Gamon, Julia A. "Similarities and Differences." *Agricultural Education Magazine* 66:11 (May 1994): 4-5.

Introduces the theme of cooperation in agricultural literacy that is addressed by numerous authors in this special issue of the magazine by Donna L. Graham and others. Compares agricultural teaching and cooperative extension to highlight similarities and differences and encourages a greater sharing of resources and ideas at local levels.

462. Gilles, Jere Lee. "Is Agricultural Extension for Women?" *Journal of Extension* 20 (March-April 1982): 10-13.

Focuses on important questions concerning the development of agricultural extension programs for women: (1) Are women sufficiently involved in agriculture to warrant special programs? (2) Do present agricultural extension programs adequately serve women? and (3) Does the design of agricultural extension programs for women violate civil rights legislation? Concludes that the development of programs for women is a new frontier for agricultural extension for which guidelines are needed; that this offers tremendous opportunities to increase agricultural production and to improve the quality of life; and that this program can be achieved most effectively by improving the productivity of an important part of the farm labor force--women.

463. Gorman, Pat. "Women and Agriculture--A Two-Year College Student's View." *Agricultural Education Magazine* 47:12 (June 1975): 280, 282.

Focuses on changes in agricultural education between 1957 and 1975. Notes significant development in content of textbooks from all references to male interest in farming careers to gender interest perspectives and expansion of the profession. Points to demand for more specialized teachers who do not necessarily need agricultural backgrounds and an increase in the number of agriculture and mechanics teachers.

464. Graham, Donna L. "Teaching and Extension-Career Paths and Interactions." *Agricultural Education Magazine* 66:11 (May 1994): 8-9.

Personal report from a woman practitioner who pursues extension education as a career. Describes her work as a county agent, as a 4-H agent, and state specialist. Points to the lack of understanding between teachers and county agents as the main barriers to the types of cooperation and networking that improve the community and cites ideas for implementation.

465. Gregg, Ted, et al. "Some Myths About Women Agriculture Teachers." *Agricultural Education Magazine* 47:12 (June 1975): 273-274.

Teaching agriculture is a new area for women, but surveys to determine the extent of women agriculture teachers in the USA show that they are becoming increasingly involved.

466. Hewitt, Mary, and David L. Howell. "Being the Odd One in a Profession." *Agricultural Education Magazine* 63:8 (February 1991): 19-20.

Discusses the entrance of women in the profession of agricultural education since 1969 when females were officially admitted to the Future Farmers of America (FFA) organization. Shows statistics on agricultural subjects currently taught by females in the eastern half of the United States: 49% teaching horticulture; 17% teaching forestry; 30% teaching agribusiness; 46% teaching animal science; 46% agricultural mechanics; and 31% in other areas of vocational agriculture. Discusses the difficulties of being a minority in the profession and recommends the establishment of support groups that have proven to be especially useful in other professions.

467. Higgins, Kathleen Mansfield. "What Kind of Training for Women Farmers?" *Convergence: An International Journal of Adult Education* 15:4 (1982): 7-18.

Summarizes a report to the Botswana Ministry of Agriculture. Evaluates a nonformal education program for women farmers provided by the Botswana Ministry of Agriculture for Women Farmers at Rural Training Centers and confirms that training services and content of courses for women tended to be wholly domestic and much less thorough than those for men. Unveils plans by the Ministry of Agriculture to open two new centers with courses for women and plans by the Government to increase emphasis on rural development. A substantial contribution to the history of Botswana's education record.

468. Hill, Frances. "Farm Women: Challenge to Scholarship." *Rural Sociologist* 1:6 (November 1991): 370-382.

Assesses the current state of knowledge of women farmers in America and suggests directions for future research as well as research design. Suggests that social scientists can learn a great deal by studying farm women about the persistent importance of kinship in industrial society, individual roles in family contexts, the issue of defining individual rights while acknowledging the continuing realities of kinship; development of a broad concept of research that includes hypothesis formulation and data collection; and regarding people as partners in research rather than targets of inquiry.

469. Hillison, John H., and Penny L. Burge. "Women Are in Agricultural Education to Stay." *Agricultural Education Magazine* 56:5 (November 1983): 18-19.

   Suggests that women have made greater strides toward full access to agricultural education than in other disciplines since 1970 as reflected in increased female enrollment in secondary, post secondary and adult programs. Similar increases are noted in the number of female agriculture teachers, educators, Future Farmers of America (FFA) membership, and officers. Recommends that future development be built on the successes of the past through stronger involvement in total agriculture programs.

470. Khalil, Ahmed A., and S. A. Ilyas. "Strategies for Sustained Advancement of Women in Rural Areas." *Journal of Rural Reconstruction* 27(1): (1994): 7-23, 1994.

   Recognizes women's role in household income and the need to sensitize policymakers about women's issues such as education, health care, and human resource development as significant factors for the advancement of women. Points to illiteracy, poor health, malnutrition, family care, and limited role in agricultural production as constraints to improvement. Cites thirteen issues which must be addressed for successful development of females.

471. Kinsey, J. "Women in Agriculture: the US Experience." Staff paper of the Department of Agricultural and Applied Economics, University of Minnesota, 1987.

   Studies the progress and status of women in the United States whose interests, education, and training enabled them to be employed in professional and technical off-farm occupations related to agriculture. Discusses the perceived educational needs of women who are either farmers or professional workers in agricultural business, education, government and research and predicts their future in agricultural careers. Shows that barriers still exist which hinder women from obtaining suitable education, jobs and wages in both rural and urban areas, agricultural firms and colleges.

472. Klepper, Betty. "Long-term Career Goals for Professional Women in Agriculture." *Journal of Agronomic Education* 15:1 (Spring 1986): 29-33.

Indicates that most women have a greater range of roles and status than men, that men's roles and status tend to be congruent to support one another, that men perform their professional work to fulfill a family role while women in the profession have conflicting roles between career and family with multiple goals. Recommends the formulation of goals early in a career to guide behavior, decisions, and priorities and give direction to future development. Women must move toward long-term goals along the same pathways that men have used, holding a series on positions with increasing responsibilities and establishing professional credentials as they progress in the profession. Women bring special strengths such as adaptability, flexibility, and good communication abilities to the agronomic workplace.

473. Kolde, Rosemary F. "Women in the Labor Force." *Vocational Education Journal* 60:7 (October 1985): 23-29.

Discusses factors that contribute to women's increased labor force participation rate. Shows rapid increase of women enrolled in agriculture, horticulture, natural resources, agricultural products, and mechanization as well as in forestry disciplines. Recommends that recruitment material include pictures of women, that nontraditional students and traditional workers participate in career day programs and conferences, that continuing efforts be made to inform students about nontraditional options, that parents be provided information which reflects the range of opportunities for women and young women in vocational education, and that work with employers be continued to obtain highly skilled workers regardless of gender.

474. Knotts, Don, and Rose Knotts. "Why So Few?" *Agricultural Education Magazine* 47:12 (June 1975): 269, 276.

Addresses reasons why there are so few women in agricultural education and points to early social conditions, occupational counseling, social attitudes, and institutional limitations as some of the major causes. Recommends that measures be taken to overcome the barriers.

475. Kuehl, R. J., et al. "Progress and Opportunities for Women in Agricultural Sciences." *Journal of Agronomic Education* 16:2 (Fall 1987): 55-60.

Reviews a five-year status report on women and career opportunities in agricultural sciences. Because enrollment of both men and women in agricultural colleges and universities declined in the 1980s a critical shortage of quality agricultural scientists developed. Although the number of women earning Ph.D.s in agricultural sciences increased substantially in the 1980s, it is still low compared to the number of doctoral degrees earned by women in other fields, and women's salaries still lag behind men's in most reports. Recommends that colleges and universities provide greater opportunities for women to develop into role models.

476. Kumar, K., and Sneh Lata Mago. "Training Needs of Farm Women in Haryana." *Indian Journal of Adult Education* 35:10 (October 1974): 72-76.

Sees women as an integral part of all farm training programs as they play a significant role in the decision-making process on farm matters, perform many of the farm operations, and undertake many responsibilities concerning care and management of farm animals. Describes a new program for women based on a study of 102 women which was designed to reveal training needs of farm women, timing of training, and duration of training and methods of training. Identifies eight educational needs of farm women.

477. Kuznik, Anthony. "Women in Agriculture in a Two-year College." *Agricultural Education Magazine* 47:12 (June 1975): 275-276.

Focuses on women majoring in agriculture at the University of Minnesota Technical College. There is a high demand for people to take agriculture positions, and this article states that women can definitely assume these responsibilities.

478. Lee, Delene W., and Shirley Haggard. "Expanding Opportunities for Women." *Agricultural Education Magazine* 58:6 (December 1985): 18-19.

Addresses opportunities for women as farm entrepreneurs and in the nonfarm sector, opportunities for vocational education in agricultural industry, and ways teachers of vocational-technical education in agriculture can facilitate desired changes in the status of women.

479. Lee, Jasper S. "Theme: Future Programs of Agricultural Education." *Agricultural Education Magazine* 58:6 (December 1985): 4-19.

Discusses factors for engendering the future, the impact of legislation and vocational-technical education in agriculture. One of a series of articles in this issue which is devoted to visions and plans for the future.

480. Leibelt, Don C. "Women in Urban Agribusiness." *Agricultural Education Magazine* 47:12 (June 1975): 285.

Indicates that between 1967 and 1975 the enrollment of young women in agribusiness classes at the high school level increased from 1 to 50. The young women have been above average scholastically and ahead of the young men in maturity.

481. Leske, Gary, et al. "Theme: Experiential Learning." *Agricultural Education Magazine* 67:3 (September 1994): 4-22.

Focuses on the relationship of experiential learning to school-to-work transition, science fair projects, leadership training, performance-based assessment, livestock judging, management instruction for farm women, and instructor preparation.

482. Lindsay, Beverly, et al. "Women and National Development in Africa." *Western Journal of Black Studies* 1:1 (March 1977): 53-58.

Examines the role of women in the development of Africa and factors influencing the integration of African women into the modern economic sectors. Provides historical and social reasons for the traditional roles held by women in agriculture. Makes recommendations for the educational development of women.

483. Manjula, N., Siddaramaiah, B.S., and Lalith Achouth. "Factors Contributing to the Adoption Level of Trained and Untrained Farm Women--An Application of Principal Component Analysis." *Journal of Rural Development* 13:1 (1994): 107-114.

Reveals education and training as basic requirements for rural development. Studies behavior factors that contribute to adoption of new practices and technology by trained and untrained farm women including such variables as marital status, education, landholding, and scientific orientation. Includes 40 trained and 40 untrained farm women who participated in a 10-day training institute dealing with groundnut cultivation practices at the University of Agricultural Science in Bangalore, India. Shows strong connection between knowledge and improvements in productivity. Shows that adoption of new technology and practices is highest where the farm woman is an elder member of the family. Recommends more follow-up on extension training that would benefit farm women.

484. Mbilinyi, Marjorie J. "Access to Education in East Africa: An Introduction." *Rural Africana* 25: (Fall 1974): 1-4.

Updates empirical findings on the problem of access to formal and informal education for farm women. Raises critical concerns about the measurement of stratification as the main determinant for educational access in underdeveloped societies in order to stimulate further investigation. Reports that access to primary education depends on the sex of the child (available to boys rather than girls); that the level of the mother's education is an important predictor of the child's access to education; that different expectations for education and work exists for boys and girls; that increasingly closed recruitment exists based on socioeconomic condition, parental occupation, and formal educational attainment; that a disproportionate number of opportunities for access to education by males and females; that primary school intake is drawn from rich and middle level peasants; that adult education provided by universities is monopolized by wage earners with relatively high levels of formal education. Confirms those trained used their education in private and self-employment, that many articles written on the topic lack clear theoretical framework, and that much of the research failed to raise the question of what type of access to what kind of education, and revealed constraints rather than open access to education in East Africa.

485. Mbilinyi, Marjorie J. "Research Priorities in Women's Studies in Eastern Africa." *Women's Studies International Forum* 7:4 (1984): 289-300.

Analyzes research on the significance of class differentiation among women and the need to distinguish different kinds of male-female relations both within and across the different classes in society. Suggests the need for research which focuses on different stages of capital accumulation process and different forms of capital accumulation as they relate to concrete struggles among classes of women. Research priorities include a critical investigation of the development of women-related studies from the colonial period forward and an examination of the role of donors in funding for women-related research.

486. Meera, B., and G.T. Nair. "Impact of Farm Women's Training on the Adoption of Improved Agricultural Practices." *Agricultural Research Journal of Kerala* 28: (1990): 57-58.

Evaluates the impact of one of the farm women's training programs conducted in Kerala (Tiruvnananthapurmam district). Involves 80 trained farm women, 40 in a controlled group, to assess improved agricultural adoption practices. There was a significant difference in the extent of adoption of improved agricultural practices in the farms of trained and untrained farm women, and the adoption level could be enhanced by training farm women.

487. Obidi, S. S. "Indigenous Agricultural Education of the Yoruba in Nigeria." *Scandinavian Journal of Development Alternatives* 10:4 (December 1991): 77-89.

Portrays males as the main practitioners of agriculture in traditional Yoruban society in contrast to the mainly female practitioners in many other countries of the continent. Outlines the nature, scope, and method of indigenous agricultural education for men which prepares them to be productive farmers. Discusses a study of 20 females and 35 male farmers based on interviews and written sources. Points to shortage of teaching personnel, inadequate machinery, and farm equipment as major challenges.

488. Okeyo, Achola Pala. "Reflections on Development Myths." *Africa Report* 26:2 (March-April 1981): 7-10.

Supports the rationale of improvement of concepts, methodologies of research, and data collection to clarify and substantiate the various manifestations of inequality between the sexes. Cites international encounters which reveal the global consensus on the diagnostic aspects of the problems facing women and the lack of consensus regarding strategies for gender equality, the continuing internationalization of the situation in the context of the United Nations, greater understanding of conditions for women in Africa, and development of methodologies to encourage openness across cultures in order to improve the status of females in societies.

489. Paret, Andrea Martha. "Women Agricultural Graduate Study: A Follow-up of Female Graduates of the College of Agriculture at Oklahoma State University from 1985-1989." *NACTA Journal* 35:1 (March 1991): 46-49.

Reports on a follow-up study of female graduates of the College of Agriculture at Oklahoma State University to determine the status of enrollment and graduates in agriculture, female graduates' perceptions of the relationship between their areas of employment and their college degrees, the adequacy of programs within the College of Agriculture as perceived by female graduates, as well as perceptions of sex bias and sex stereotyping in their college training and work. Shows moderate benefits of degrees, satisfaction with the choices of study; that being married had a positive impact on their careers; and that most were treated differently from men in employment. Suggests that more attention be placed on job placement, the use of women as role models, and further research in this area to compare female and male graduates' expectations regarding job placement services and future employment, and methods used.

490. Quisumbing, A. A. "Intergenerational Transfer in Philippine Rice Villages: Gender Difference in Traditional Inheritance Customs." *Journal of Development Economics* 43:2 (1994): 167-195.

Examines education, land, and nonland asset transfer from parents to children in five rice villages in the Philippines in order to maximize wealth or equalize sibling incomes or tradeoffs between these two options. Focuses on gender differences or preferences of parents in inheritance decisions and tests the importance of family-specific unobserveables against individual heterogeneity as a determinant of transfers as well as showing how differential effects of parental endowments differ from common preference models of intergenerational transfers. Uses data from a retrospective survey that analyzes intra-household differences in transfers in these villages. Provides rare information on intergenerational transfers in rural economies including human and physical wealth of parents at the time of marriage, characteristics of children, and specific forms of transfers. Found that wealth constraints affect parents' educational investment in their children, that poor families tend to concentrate educational investment in the oldest child and the benefits from the secular expansion of educational opportunities, that daughters of better educated mothers receive more land, nonland assets, and total inheritance while better educated fathers give land preferentially to sons but favor daughters in education.

491. Rai, Shirin M., and Zhang Junzuo. "Competing and Learning: Women and the State in Contemporary Rural Mainland China." *Issues and Studies* 30 (March 1994): 51-66.

"Competing and Learning" is a campaign aimed at moving the country beyond stagnation in the production of food grain, poultry, and pigs by increasing levels of agricultural production, increasing literacy levels of girls and women in rural areas which have a 70% illiteracy rate, and improving women's social status. Sponsored by the All-China Women's Federation in collaboration with twelve other organizations, this project targeted women as the major food producers and sought to address the question of their productivity. Explores themes that are beginning to emerge from this project (it is still in progress in specific regions (*huo-tung*) of China). Notes that the issue of women's labor in agriculture had been "upgraded"(p. 56). The authors recommend teaching rural women to read, write, and learn farming skills and technology and to compete for achievements and contributions to production. They expect to enrich education of 60 million women in each five-year project and support the overall goal of modernization in mainland China.

492. Ram, Rati, and Ram D. Singh. "Farm Households in Rural Burkina Faso: Some Evidence on Allocative and Direct Return to Schooling, Male-Female Labor Productivity Differentials." *World Development* 16:3 (1988): 419-424.

Assesses the allocative and direct components of the productivity effect of schooling and the male-female productivity differentials in the setting of a highly traditional farming system in Burkina Faso. Shows a 10% rate of return to schooling based on the direct effect for family members other than the household heads and, for the head, the allocative component is higher than the direct effect. Reveals a higher productivity of labor input for women.

493. Rea, Jennette, et al. "Lessons from Women in the Agricultural Sciences." *Agricultural Education Magazine* 62:2 (August 1989): 20-21.

An examination of the lives and careers of women scientists as well as the diverse interest, significant contributions and challenges revealed in the study of these women. Scientists studied include Elizabeth Pickney, Jane Colden, Harriet Williams Russell Strong, Anna Botsford Comstock, Beatrix Potter, Alice Catherine Evans, Edith Marion Patch.

494. Reynolds, Carl L. "Should We Encourage Women to Enter Ag. Ed.?" *Agricultural Education Magazine* 47:12 (June 1975): 272, 274.

Reports on a survey of enrollment of students in the member associations of State Institutions of the National Association of State Universities and Land Grant Colleges in 1974. Women represented 23% of total enrollment with 13.6% increase in enrollment, of which only 12% were registered in teacher training in agricultural occupations. Showed low interest in agricultural economics, agronomy, and agricultural mechanization. Recommends encouraging women to enroll in agricultural curricula, facilitating expansions to accommodate students from urban settings, and taking advantage of the fact that women relate better to younger students.

495. *Rural Women's Participation in Development.* N.Y.: United Nations Development Programme, June 1980.

Devotes several chapters to rural women's education and training. Points to education and training as the undisputed major factors that promote national development and positive change. Identifies some of the reasons for disparities between men and women in education and technical and vocational training as well as policies and programs of the United National Agencies in Africa, Asia, Pacific, and Latin America. Identifies major challenges for the future: eradication of illiteracy, universal provision of elementary education, training and implementation of new technology and vocational skills.

496. Schwieder, Dorothy. "Education and Change in the Lives of Iowa Farm Women, 1900-1940." *Agricultural History* 60:2 (Spring 1986): 200-215.

Gives history of the development of practical education for Iowa farm women from 1900, when they had access to both county and state farm institutes, which consisted of short courses or lectures by professors from Iowa State College. Discusses educational materials included in *Wallaces' Farm* journal and courses on home economics. Reflects much progress in practical education for farm women by the end of the 1930s, including some leadership roles for women.

497. Seevers, Brenda. "Preparing Agriculture Teachers and Extension Agents." *Agricultural Education Magazine* 66:11 (May 1994): 6-7.

Recommends restructuring Agricultural and Extension Education programs and emphasis on teacher preparation to meet new challenges for the year 2000 and the linking of theory with practice via direct field experience, community-based education, human resources development, and work with diverse populations.

498. Sibalwa, David. "Women's Contribution Toward National Development in Zambia." *Convergence: An International Journal of Adult Education* 26:2 (1993): 38-46.

Focuses on one of Zambia's major problems--the lack of recognition of women's significant traditional contributions toward national development and the general quality of life. Shows that women are mainly responsible for 70% of subsistence production as men handle cash crops and migrate to cities in search of jobs. Cites legislative constraints that sanction subordinate roles for women. Reveals recent interest and efforts to enhance women's productivity by providing greater access to education, skills training, employment, and pay equity. Urges the implementation of more formal and informal education for women that take into account the multiplicity of roles. Affirms that a trained woman is a trained family. Education for all people in Zambia has helped to narrow the gap between men and women in jobs.

499. Smith, Suzanna, and Barbara Taylor. "Curriculum Planning for Women and Agricultural Households: the Case of Cameroon." in Gladwin, Christina H., *Structural Adjustment and African Women Farmers*, Florida: University of Florida Press, 1991, pp. 373-386.

Recognizes the fundamental roles played by women in increasing food production and promoting self-sufficiency and urges developers to provide agricultural training and extension services for farm women. Evaluates a course created and planned by participants at the University Center of Dschang, Cameroon's agricultural training institution. Assesses components of the course and indicates that this format is an effective vehicle for initiating the first steps in curriculum planning for the women and agricultural households program at the institution.

500. Sproles, Elizabeth Kendall. "Perceptions by Nontraditional and Traditional Agricultural Students toward Their High School Preparation and Work Barriers." *Journal of the American Association of Teacher Educators in Agriculture* 28:2 (Summer 1987): 18-24.

Shows an increase in female enrollment in vocational and agricultural programs as a result of federal legislation which mandated access to women in areas other than the traditional home economics programs. Analyzes the attitudes of nontraditional and traditional program completers toward their school preparation, work, and barriers encountered in obtaining employment after completing their vocational agriculture program. Provides data that can be used to assess strengths and weaknesses in programs as well as to identify motivators to stimulate increased enrollment. Identifies employment as the area of greatest problems for nontraditional program completers in agriculture. Expresses the need for special efforts to insure that students are successful in educational programs and careers which are considered nontraditional.

501. Standing, Guy. "Education and Female Participation in the Labour Force." *International Labour Review* 114:3 (November-December 1976) :281-297.

Discusses equity issues in the education of women and young women in industrializing countries as a growing proportion of national income is being allocated for education based on an awareness that human investment is the most effective means for stimulating economic growth. Concludes that the spread of education tends to create a social climate favoring female economic activity and lowering barriers to their employment in higher-income occupations.

502. Steele, Roger E., Comp. *Association for International Agricultural and Extension Education Conference Papers.* 10th Annual Conference, Arlington, Virginia, March 24-26, 1994. ED368914.

Presents selected papers focusing on international aspects of agricultural and extension education, some of which focus on issues specifically related to the education of farm women. Provides useful information about agricultural education for women.

503. Stephens, Alexandra. "Papua New Guinea: Gender Roles and Agricultural Education." *Agricultural Information Development Bulletin* 13:3 (September 1991): 12-14.

Charts the history of the innovative "Rural Life Development" curriculum in Papua New Guinea's agricultural colleges. The program failed as it did not take into account the fact that women were the major producers of food. Courses for women were confined to home economics and child care. Although not successful, the program stimulated discussion that identified administrators who were more sensitive to the real needs of women farmers and who would provide educational strategies for working with political communities.

504. Thomas, John K., et al. "Industry Notes Career Development of Agricultural Graduates: A Gender Comparison." *Agribusiness* 7:5 (1991): 503-514.

Examines the employment mobility of male and female agricultural graduates. Findings of a mail survey to agricultural alumni respondents showed that female graduates were less likely than male graduates to have entered agricultural jobs after completing college and to have received lower salaries for comparable employment. Revealed that one-fourth of the women became homemakers, employed part-time, or unemployed, producing a net loss of agricultural capital.

505. Thompson, Barbara. "Ag Division Looks Ahead with Confidence and Concern." *American Vocational Journal* 52:2 (1977): 32-37.

Discusses the role and status of vocational, agricultural, and agribusiness education in the context of the total educational process. Points out some directions for future planning to improve agricultural education. Includes comments about exports, training programs, curriculum development, teaching teachers, continuing education, constructive dialogue, as well as the importance of Future Farmers of America (FFA) organization and Friends.

506. Thompson, O. E. and L. Z. McCandless. "Women in Agriculture: The New Growth in Programs." *Agricultural Education Magazine* 52:7 (January 1980): 19-20.

Shows the tremendous impact of the 1968 Supreme Court ruling which mandated the admission of females into the previously male-only FFA (Future Farmers of America) organization. By 1980, females made up one-third of the over 100,000 Californians studying agriculture and similar developments were taking place throughout the nation. Shows the shift in curricula concentration from animal science and ornamental horticulture to agricultural economics, resource sciences, and business aspects of agriculture, the impact of which will provide integration issues for educators and employers.

507. "Training 8 Million Farmers by 2000." *Beijing Review* (May 24-30 1993):6.

A short projection of how China plans to teach 8 million farmers new skills by the year 2000. Between 1993 and 1995, 500 countries and 3 million farmers, a significant portion of which are women, are expected to participate in the "green certificate" program designed to train two or three farmers in every village in advanced techniques which they will share with others in their village. It is stated that the course becomes popular after newly trained farmers earned more money with their new skills. The pilot project has been in operation since 1990, in which 40,000 farmers have been awarded certificates upon the completion of their training.

508. Vetter, Louise. "Where the Women Are Enrolled." *Vocational Education Journal* 60:7 (October 1985): 26-29.

Reviews the status of enrollees a decade following Title IX Education Amendment legislation which opened access to agricultural education for women. Reflects most profound increases for women in agriculture from 5.4% in 1971-72 to 31.5% in 1981. Suggests strategies for broadening the range of nontraditional opportunities for women by using statements, photographs to show that women are welcome in all vocational education programs, use of nontraditional and traditional students in panel presentations and career day conferences, media publicity and parent-school organizations, and work with employers to obtain highly skilled workers regardless of gender.

509. Vlassoff, Carol. "From Rags to Riches: The Impact of Rural Development on Women's Status in an Indian Village." *World Development* 22:5 (1994): 707-719.

Discusses change in women's status in a village in Maharashtra that experienced rapid agricultural growth during 1975-1987. Examines two objective indicators of women's status--education and mobility--as well as attitudinal indicators of autonomy, economic power, and prestige. Economic growth is found to have a mixed effect on women's status. While freeing women from much tiresome work and giving them more prestige within the home, it resulted in reduced autonomy and economic power. Seeks explanations in modern theory on women and development and in historical evidence concerning the impact of industrialization on women.

510. Whittington, Susie. "Retaining Women in Nontraditional Employment." *Agricultural Education Magazine* 63:2 (August 1990):18-19,23.

Relates extensive increase in females in the U.S. labor force since 1974 in response to economic demands of operating a household. Reports on a study which suggested that agricultural education was a profession that offered a place for women and a corresponding increase in women student enrollment since then. Reflects traditional attitudes regarding care-giving, technical competence, professional involvement, and gender issues that hinder the profession from establishing a longevity record for women. Recommends that those in the profession assume responsibilities to encourage the retention of competent women in agricultural education.

511. Zambia Association for Research and Development (ZARD). "Agenda for the Study of Rural Women in Zambia: ZARD's Presentation at Forum '85." *International Studies Notes* 14:3 (Fall 1989): 72-76.

Shows that women receive less formal and nonformal agricultural education than men even though women are the major farm producers. Women are excluded from access to essential techniques for improving their productivity as farmers. Since most of the women are illiterate, they cannot learn from the pamphlets distributed by the Extension workers. They have limited access to modern technology and technical skills that provide labor-saving devices to improve efficiency. Male extension workers do not consider women as part of their clientele. Even though the United Nations Decade of Women had motivated some interest, special projects have not resulted in the integration of women into overall education programs. Shows that Zambia does not yet have a national action research program relevant to improving the education of rural women. Makes recommendations for such development.

DISSERTATIONS AND MASTER'S THESES

Reference librarians in academic institutions often receive inquiries for dissertations on specific subjects. University of Michigan Microfilm International (UMI) provides access to all American dissertations accepted at accredited institutions since 1861. Canadian dissertations and an increasing number of dissertations from abroad are also included. The subject of "women farmers" or "women and agriculture" was searched in this database to collect information on related degrees. Annotations are included for some of the items included in this chapter.

512. Abbott, Susan. *Full-time Farmers and Week-end Wives: Change and Stress among Rural Kikuyu Women.* Ph.D., The University of North Carolina at Chapel Hill, 1974.

513. Adams, Elizabeth Rose. *Women Primatologists in National Parks and Equivalent Reserves in East Africa.* Ph.D., Clemson University, 1989.

    The role of women scientists in parks was examined through an assessment of women's scientific productivity in national parks and equivalent reserves. Empirical evidence was sought to determine the existence or non-existence of a "female" perspective in the field of natural science.

    Primatological field reports by women in the national parks and equivalent reserves of East Africa were collected from the scientific literature. The reports were matched with companion pieces, field reports by men published concurrently in the same journals. The reports were sampled and analyzed for vocabulary reflecting gender and concepts of cooperation and competition, connection, and separation. A discriminant analysis was performed on use of nine variables by three gender groups, male, female, and mixed authorship.

    Results revealed a statistically significant composite variable on which each report was measured. Major differences were found in use of female gender words. Male authors use male gender words twice as often as they do female gender words, whereas female authors use both equally. Female authors can also be identified by their use of cooperative words such as "affiliation" or "bonding." Male authors show a heavier use of competitive words such as "adversary" or "struggle" than do female authors.

Parks and reserves are acknowledged as valuable open-air laboratories where the study of free-ranging animals is possible. One possible explanation of women's success as field naturalists is that parks offer an atmosphere of freedom allowing for studies that diverge from orthodox science.

514. Alberti, Amalia Margherita. *Gender, Ethnicity, and Resource Control in the Andean Highlands of Ecuador.* Ph.D., Stanford University, 1986.

Explores how ethnicity and resource control interact to affect the roles and status of women and men in the Andean Highlands of Ecuador. Using a case study format, examines the thesis that socioeconomic differences have minimal impact on the gender roles and status of women in societies that culturally accord them a productive economic role in the household.

The fieldwork took place in three adjacent, but ecologically and ethnically distinct, communities in the central Andean highlands. The households in one community were overwhelmingly mestizo, where the norms demand female economic dependency. In another community the households were indigenous, where the norms extend an economic role to women at the household level. The third community was mixed.

The methodology employed integrated ethnographic techniques with quantitative analysis. Data collection included participant observation, a time allocation study, and a household survey in which male and female household heads of 67 households were interviewed. Patterns of household labor use, involvement in decisionmaking, and the ability to allocate household primary resources were examined. The relevance of development programs of sensitivity to ethnic as well as socioeconomic variability was underscored.

515. Alexander-Flinn, Gretcha Evon. *Perception of Decision Makers Whose Programs Impact Rural Sector Women in Less Developed Countries.* Ph.D., Texas A&M University, 1991.

To determine the perceptions of selected decision makers concerning gender issues related to international rural sector development programs, questionnaires designed to answer ten research questions were mailed to 300 decision makers who were participating actively in international development activities at the time of the study. One hundred and eighty questionnaires were returned. Findings suggest that there were no significant differences among the perceptions of individuals employed by governmental organizations, non-governmental organizations, private voluntary organizations, university research programs, and university administration concerning gender issues. The decision makers held generally positive opinions about working with women's programs. They recognized the importance of gender issues, and their perception was that only 17% of women's needs are being met by current international projects.

516. Alston, Margaret Mary. *A Study of Farm Women: An Australian Feminist Perspective*. Ph.D., University of New South Wales, Australia, 1993.

Examines the lives and experiences of Australian farm women from a feminist perspective. Focuses on the economic contribution women make to ensure the survival of family farming, the economic, technological and cultural changes affecting farm families and how these changes effect women's roles on the farms, and also how those changes have affected farm women's traditional conservatism and avoidance of public roles. A qualitative methodological framework is employed, where semi-structured interviews were conducted with 64 farm women in New South Wales in 1991. Thirty-two women live in predominantly wheat growing areas, and thirty-two women live in livestock grazing areas.

It is apparent from this study that women are making a major economic contribution to their enterprise through their unpaid family work, on-farm work, off farm work, and community work. The economic downturn has led to women replacing hired labor and moving off the farm to seek work. Technological changes have pushed women to the periphery of production despite the fact that many women drive machinery. These communities accept now that women will work off farm and will seek higher education. A significant shift in attitude is occurring among farm women who are no longer prepared to tolerate lack of recognition for their efforts. As they take up a more visible position in the economic and political life of their communities, they are seeking acknowledgment.

517. Angeles-Bernardo, Estelita. *Motivational Factors Influencing the Vocational Choice of Home Economics and Agriculture as a College Major by Women of the Central Luzon State University (Philippines)*. Ed.D., Oklahoma State University, 1986.

Compares motivational factors in student choices of home economics (traditional) and agriculture (non-traditional) curricula. The sample was taken at random from each of the curriculum group rosters provided by the Central Luzon State University registrar's office. There were no significant differences in the ages nor in the places of residence of the traditional and the non-traditional groups. However, there was a notable difference in the college year of students in both curriculum groups in which a larger percentage were found in the senior year. A significant difference was shown in class levels as students in both groups chose their careers at their sophomore year. There were no significant differences in the self-rating of the probabilities for employment after graduation.

518. Anstey, Barbara Eleanor. *Pesticides and Iowa Farm Women: Knowledge Levels, Information Sources and Educational Needs*. Ph.D., The University of Iowa, 1983.

Examines the knowledge of insecticide safety of Iowa farm women residing on farms where corn insecticides are used. Looks also at the sources women used to learn about the safe use of corn insecticide and to discover if any association existed between the level of knowledge and selected factors including geographical location of the subjects and residency. Personal interviews took place with 165 Iowa farm women. The questions were drawn up on the basis of an extension literature review and interviews with chemical dealers, County Extension directors, farmers, farm women and health authorities. The results revealed that while there was widespread anxiety about possible dangers involved with insecticides, most of the women interviewed were not fully informed about insecticide usage. Despite the fact that farm women often have direct involvement with farm work using insecticides, there are at present few information channels which are directly aimed at providing them with insecticide safety knowledge. Their current sources of information are indirect, limited, often contradictory. New programs need to be developed which utilize Cooperative Extension Services and employ adult education principles in order to promote a dramatic increase in women's learning about insecticide safety and handling practices.

519. Barton-Cayton, Amy Elizabeth. "*A Women's Resistance is Never Done*": *The Case of Women Farmworkers in California.* Ph.D., University of California, Santa Cruz, 1988.

Explores the possibility that women farmworkers are so situated in social reality that their resistance struggle is structurally predisposed to being non-reformist in nature since it simultaneously confronts both capitalism and patriarchy. Survey data gathered in a 1977-78 sociodemographic study is used from 600 men and women farmworkers. Participant observation and in-depth interviews conducted between 1973 and 1980 further inform the dissertation. The primary source material and the literature on farm labor illuminate the structure of California agricultural production, the location of women farmworkers at the workplace and in the family, and the types of farm labor resistance. Interview data and the literature on farm labor also indicate that patriarchal and racist beliefs of agricultural employers are a significant part of the basis of discriminatory job segmentation in California agriculture. The conclusions suggest specific characteristics of resistance strategies for social change, not just for farmworkers, but for any effort which confronts the capitalist system.

520. Baser, Heather Jane. *Lima and Women Farmers in Zambia.* M.A., Carlton University, Canada, 1988.

The basis of the message disseminated by the Zambian agriculture extension services is a series of standardized measures for the application of seeds and fertilizers based on a lima measure of land or one quarter of a hectare. This thesis looks at the impact of the so called lima recommendations on women farmers using Norman Girvan's three categories of technology: organization, production, and consumption. The conclusion is that all three categories of the technology are inappropriate for most women. The consumption technology, primarily hybrid maize, requires a production technology (information and physical resources) to which women seldom have access because of the ownership structure and the social relations of production of the society (organizational technology). Greater access to the resources required for increased productivity will only come with improved social and legal status for women. Effecting changes in this status is thus a crucial development issue.

521. Berkowitz, Alan David. *Role Conflict in Farm Women.* Ph.D., Cornell University, 1981.

The causes and effects of role conflict were studied in 123 randomly selected dairy farm wives in four New York counties. The farm wives completed a Likert-style questionnaire containing measures of husband's supportiveness, husband's home involvement, wife's farm involvement, farm style, age, number of children, development stage, education, and outside employment. The variables were grouped into four factors: a role-related factor, a demographic factor, a family development factor, and a farm-related factor. The questionnaire also measured conflict between farm and home responsibilities (inter-role conflict) and conflict between husband and wife over the wife's role (interpersonal role conflict).

Inter-role conflict was negative related to husband's supportiveness, development stage of the family, and age of farm wife. Interpersonal role conflict and wife's farm involvement were positively related to inter-role conflict. Interpersonal role conflict was negatively correlated with husband supportiveness.

The results, when compared with research conducted with non-farm women, indicate a number of similarities: role conflict with both sets of women is associated with age, development stage of the family, and number of children. The findings were interpreted as confirming the importance of the wife's farm role and the role of the husband-wife relationship in determining farm success or failure.

522. Borish, Linda Jane. *"The Lass of the Farm": Health, Domestic Roles, and the Culture of Farm Women in Hartford, Connecticut, 1820-1870.* Ph.D., University of Maryland, 1990.

Examines the health of farm women in Hartford County, Connecticut, and the debate in the rural community. Indicates that issues of health and physical well being functioned as part of a larger male-female conflict over rural life. Women and men exercised different levels of power and they differed in their interpretation of the nature of farm living along gender lines. The exposure of young rural women to city culture and new options and their ensuing farm exodus shocked female agriculturists and some reformist males who agreed with them. They asserted that farm females' terrible quality of life needed to be redressed and promoted physical recreation and sport, improved mental culture and home embellishments, domestic labor and health-saving devices, and participation at agricultural fairs as antidotes to the perceived stagnation and ill health of the rural women. Rural reformers believed they could remedy farm life and satisfy women's increased expectations.

523. Brown, Katrina. *Women's Farming Groups in a Semi-Arid Region of Kenya: A Case Study of Tharaka Division, Meru District.* Ph.D., University of Nottingham, United Kingdom, 1990.

Examines how far women's farming groups are able to foster self-reliance among peasant farmers in Tharaka Division, a semi-arid region of Kenya. It is an impoverished, drought prone region and the population pressure is resulting in intensified land use. Many households are headed by women, and the majority of farms are managed by women. Three aspects of women's farming groups were investigated: participation, extension and innovation, and access to development resources.

A comparison is made between the economic and social status of participants and non-participants in women's farming groups. If it is the case that poor women are excluded from these groups, then a policy of targeting agricultural services and inputs to women's groups actually discriminates against resource-poor farmers.

The study concludes that women's farming groups have the potential to foster self-reliance among peasant farmers. However, at present poor women do not join groups because of severe time constraints created by competing labor demands. Present policy is biased in favor of groups from more fertile areas. It is necessary to formulate policy appropriate to dryland areas where women's farming groups may provide a valuable mechanism for reducing vulnerability and ameliorating the effects of drought and famine.

524. Canoves Valiente, Gemma. *Geography and Gender: Women, Work and Family Farm.* Geog. D., Universitat Autonoma de Barcelona, Spain, 1990.

Gives a general picture of women's role in family farming and shows how women's work is underestimated. Farm women produce more goods and services for household consumption than do women from non-farm households and in this way make an essential contribution to the survival of the farm. It is stressed that though statistical analysis is necessary to evaluate women's contribution to agriculture in a comprehensive and general way, it should be complemented by fieldwork.

525. Carbert, Louise I. *Agrarian Feminism: The Politicization of Ontario Farm Women.* Ph.D., York University, Canada, 1991.

Examines three bodies of literature, liberal modernization theory and new social movements, agrarian politics, and the history of the women's movement and argues for beliefs and practices of agrarian feminism which are conservative in content but receptive to feminist consciousness.
Data were gathered from 117 in-person surveys in Huron and Grey counties in Ontario in 1989. Carbert questioned farm women about their involvement in agricultural production and their political salience, participation, conceptualization, and efficiency. "Postmaterialist" and feminist questions were posed to a population of whom such questions are rarely asked. Using standard empirical indicators of mass politics, quantitative procedures were performed to compare farm women to predominantly urban populations and to assess the impact of agricultural work variables on orientations to politics and feminism in the survey population. Analyzes the political struggles and potentials of farm women, individually and collectively, in the context of petty bourgeois, household-based agricultural production.

526. Cashman, Kristin. *A Grounded Theory Describing Factors in the Adoption Process of Alley Farming Technology by Yoruba Women in Nigeria.* Ph.D., Iowa State University, 1990.

Aiming to discover a theory from data on rural Yoruba women in southwestern Nigeria, this study deals with farmers exposed to an agroforestry technology called "alley farming." Data were collected over four years, 1984-1986 and 1988. Participant observation and open-ended interviews were conducted. A theory of agricultural change was developed to provide a framework for alley farming research and extension. From the accompanying coding and data analysis during 1984-1986, a conceptual framework emerged which corresponded with the Concerns-Based Adoption Model (CBAM) developed by Hall, Wallace, and Dorcett (1973). This model was modified to illuminate the developmental processes that farmers experienced as they implemented the alley farming technology. A final round of data collection in 1988 solicited farmers' opinions and reactions to alley farming.

Several factors that inhibit or facilitate the diffusion of alley farming were identified including clarification of Yoruba women's role in farming, crucial, but less visible, reasons for specifically targeting women in alley farming outreach, socio-cultural conflicts and congruence factors, and power exerted from outside the cultural system.

527. Cloud, Kathleen O'Donnel. *Gender Equity and Efficiency in AID's Agricultural Projects: A Study of Policy Implementation.* Ph.D. Thesis, Graduate School of Education, Harvard University, 1986.

Examines the interaction between women's roles and the execution of agricultural projects. Is set in context of the international effort to accomplish development with equity for women as well as for men. Is formative in intent, addressing both what has been done and what has been learned so that future practice can be improved. The major questions of the evaluation include the following:
 1. Defining policy implementation as gender equitable and efficient access to project resources, what is the nature of AID's implementation experience?
 2. What are the consequences of current resource allocation patterns?
 3. What lessons have been learned that will improve project design and implementation?
 4. To what extent does examination of the projects support the argument of AID's women in development policy paper, that ignoring women's role leads to wasted resources and diminished returns on development investments?

528. Cram, Barbara Jean. *Women Organizers in the United Farm Workers: Their Motivation and Perceptions of Sexism within the Union.* M.A., California State University, Fullerton, 1981.

529. Dollahite, David Curtis. *Familial Support Resources and Individual Coping Efforts in Economically Stressed Farm Men and Women.* Ph.D., University of Minnesota, 1988.

Examines relationships between familial support resources and individual coping efforts in 270 farm men and 282 farm women experiencing economic stress. The sample was selected from Minnesota families who had completed Mandatory Farm Credit Mediation and data were collected by mail survey. Canonical correlation analyses were used to determine the relations between individual coping efforts and familial support resources. Results of analyses supported several hypotheses which stated that the relationship between familial support resources and coping efforts would differ for men and women; respondents at high and low levels of affective well-being; respondents at high and low levels perceived economic well-being; positive relationships between familial support and problem avoidance, seeking assistance, passive acceptance, physical diversion, and blaming; and positive relationships between familial support and adaptive management efforts of respondents at both high and low levels of affective and perceived economic well being.

Item loadings on canonical varieties indicated that familial support resources and coping efforts which were most highly related differed for men and women.

530. Eghan, Felicia Rosaline. *The Potential of Home Economics Personnel in Ghana to Address Development Issues Affecting Women with the Use of a Contemporary, Problem-Solving Approach in Curriculum*. Ph.D., The Pennsylvania State University, 1987.

Hypotheses were tested to determine if a relationship existed between the curriculum beliefs and perceptions of issues of women in development and selected demographic variables. Data collected used a three-section questionnaire, and 192 Ghanian home economists participated in the study. The findings indicate that almost all (99%) of the home economists were interested in gaining more knowledge and insight in agricultural production and utilization, women in development issues, and a problem-solving approach in curriculum. Significant relationships were found between curriculum beliefs of the home economists grouped by variables of education level, age, years of experience, primary professional responsibility, and country of education.

531. Endeley, Joyce Bayande Mbongo. *Women Farmers' Perceptions of the Economic Problems Influencing Their Productivity in Agricultural Systems: Meme Division of the Southwest Province, Cameroon.* Ph.D., The Ohio State University, 1987.

    Describes women farmers' perceptions of the economic problems that influence their productivity in agriculture, farm and non-farm activities, and personal characteristics in seven villages of Meme Division, Southwest Province of Cameroon. A structured questionnaire was developed and used in collecting data from 295 randomly selected women farmers. The face-to-face interview technique was employed in collecting data. Analysis of the data indicate that women farmers perceived that they have sufficient access to land, insufficient access to labor and infrastructural facilities, and insufficient access to improved farm tools, agricultural production inputs, agricultural training programs, credit, and labor-saving household technology. In decreasing order of influence on women farmers' productivity, the difficulties in accessing the following production resources were ranked from 1 to 10: improved farm tools, credit, agricultural production inputs, labor, transportation, household technology, storage facilities, land, market, and agricultural training programs.

532. Fassinger, Polly Ann. *Dimensions of Women's Work Involvement on Family Farms: A Case Study of Two Mid-Michigan Townships.* M.A., Michigan State University, 1980.

533. Frombach, Hannelore. *Stress in Farm Women: A Multivariate Approach.* M.A., The University of Regina, Canada, 1992.

    A survey approach was used to identify stress, stress symptom, depression, social support, and self-esteem levels in a sample of Saskatchewan farm women. In comparison to an earlier study by the same researcher, participants (N=191) were found to be experiencing greater levels of stress and similar levels of depression. Stressors associated with financial strains and role overload figured prominently. Farm women exhibited high self-esteem with respect to perceived competence and low self-esteem with regard to perceived evaluation by others. Demographic variables of age, employment status, type of farm, size of farm, and education level were found to be significantly related to the various stress process variables. Possible reasons for these findings are offered.

534. Gielgud, Judy. *Nineteenth Century Farm Women in Northhumberland and Cumbria: The Neglected Workforce.* D.Phil., University of Sussex, United Kingdom, 1993.

Addresses the contribution made by women to the enterprise of farming during the nineteenth century in their varied capacities as farm servants, day laborers, bondagers, farmers' wives and daughters, and as women farmers in their own right. The variety of work done by women is explored in detail and re-evaluated, supported by Day Labour Records and the reports of contemporary commentators and further interpreted by the use of specially recorded oral sources. The generally accepted decline of women's agricultural work throughout the century is challenged and evidence brought forward to support the view that it continued to be vital into the twentieth century. The major source of information on the work of women in agriculture, the two Government Enquiries in 1843 and 1867-8, are critically examined and some of the findings questioned. The author argues that the contribution of women to the agrarian economy has been seriously undervalued, to the detriment of history as a record of the past.

535. Heil, Mark Takeo. *Women and Cooperative Farming in Nicaragua: Agrarian Change and Revolutionary Transformation.* M.A., The American University, 1989.

Interviews with male and female members of five cooperative farms test the hypotheses that traditional attitudes toward the proper role of women and the quantity of daily work responsibilities borne by women restrict their ability to join into cooperative farming. Results indicate that when both conditions exist, they can prevent women from joining cooperatives. When farm leaders are favorable toward women, opportunities for women improve considerably. Women must first free themselves of the full burden of housework before they can enter into full-time agricultural work.

536. Hempstead, Katherine Ann. *Agricultural Changes and the Rural Problem: Farm Women and the Country Life Movement.* Ph.D., University of Pennsylvania, 1992.

Estimates of rural to urban migration in the early twentieth century essentially validated the Country Life reformers' concerns about rising rural outmigration. In their attempt to retain the quantity and improve the quality of the rural population, Country Life reformers devoted much attention to identifying and proposing solutions to the various social and economic dimensions of the overall "rural problem." Ultimately they were able to influence rural outmigration, and this problem no longer became a concern after the First World War. Agricultural extension education for women was greatly influenced by this movement, and more emphasis was put on practical concerns. Therefore, although the influence of the Country Life perspective ultimately waned, their initial goals and beliefs continued to inform the way in which extension work for women was justified, if not its actual content.

537. Jellison, Katherine Kay. *Entitled to Power: Farm Women and Technology, 1913-1963*. Ph.D., The University of Iowa, 1991.

Examines the mechanization and rationalization of agriculture from the perspective of rural women. Looks at the process by which farm women mechanized their work and adapted it to the standards of modern agribusiness. Rather than accept the full-time homemaker role that prescriptive literature urged upon these women, twentieth-century farm women maintained a productive role on the farm, one which relied on the use of modern technology and allowed them to assert a modicum of economic and political power within the rural community. Sources employed include prescriptive literature published by the Department of Agriculture, the Extension Services and farm life periodicals; farm women's letters and memoirs; interviews with farm women; federal census data; and visual images of farm women and technology, including advertising illustrations and photographs from the Farm Security Administration and the Bureau of Agricultural Economics.

538. Jordan, Stephen A. *The Impact of Economic Stress on Health Functioning Among Farm Men and Women as Mediated by Individual and Social Processes*. Ph.D., The University of Nebraska, Lincoln, 1989.

Investigates the relationship of the "farm crisis" to mental and physical health in farm couples. A stress model was used to identify individual, familial, and social support characteristics which differentiated individuals in their reported health system and to develop a model which could be used to understand the processes that relate financial stress to health. Forty-one farm couples completed questionnaires and interviews. Both men and women reported higher mean levels of mental and physical health symptoms than normative populations. Multiple regression procedures revealed that financial and social support measures were not predictive of health. However, subjective stressors, coping styles and family environment measures were associated with health.

Different models were necessary for women and for men. Individual coping processes were better predictors of health for men. Stress predicted reduced social support for the women. Subjective stressors and family relationships were strong predictors of health for both men and women. The data highlight the level of distress experienced by farm couples as well as the importance of farm family relationships in the moderation of negative health outcomes under stressful conditions.

539. Kleinegger, Christine Catherine. *Out of the Barns and into the Kitchen: Farm Women's Domestic Labor, World War I to World War II*. Ph.D., State University of New York at Binghamton, 1986.

Examines the changes in the domestic labor of farm women in the United States between the two World Wars. Draws largely from the prescriptive literature and letters to the editor of *The Farmer's Wife*, a popular magazine for farm women.

Farm women's household production was still an integral part of the farm economy before the twentieth century, when farm women oversaw the dairy and poultry production and raised garden produce. In the twentieth century much of this production was removed from the home, and female supervision of dairying, poultry production and truck farming were organized as agribusinesses. The farm women had no access to the increased mechanization, specialization, capital outlay, and scientific expertise that developed.

At the same time farm women were relinquishing their productive roles, the rise of consumerism provided a new role for women. *The Farmer's Wife* was originally more a primer for production with its many columns of dairy and poultry production but gradually put more emphasis on consumerism. The magazine established a Test Kitchen and Seal of Approval to endorse new products and organized a Farm Family Test Group. The sexual division of labor on the farm was redefined and reinforced by the prescriptive literature in *The Farmer's Wife*.

540. Lawrence, Roger Lee. *Implications of Characteristics and Attitudes of Farm and Village Women for Home Economics Extension Programs*. Ph.D., Iowa State University, 1958.

541. Leckie, Gloria Jean. *Female Farm Operators, Gender Relations and the Restructuring Canadian Agricultural System*. Ph.D., The University of Western Ontario, Canada, 1991.

Bridging the gap through a combined methodology, this thesis draws upon untabulated census data from 1971, 1981, and 1986 to provide background information on female farm operators in Canada. Part two of the study explores the everyday world of female farmers through personal interviews.

It was determined that female farmers were a much more diverse group than male farmers. The main concerns of female farmers were often to get access to the type of agricultural resources that male farmers take for granted. It was also felt that gender relations of agriculture are maintained and perpetuated through myth-making concerning technology, labor, and strength, which directly affected how female farmers were viewed by others in the community and how they view themselves and their work.

542. Lopez-Trevino, Maria Elena. *A Radio Model: A Community Strategy to Address the Problems and Needs of Mexican-American Women Farmworkers*. M.S. California State University, Long Beach, 1989.

Needs assessment was conducted to evaluate what Mexican American women farmworkers considered to be their major problems and needs and what relevant topics they considered important to be presented in a community-based educational radio program. Sixty women farmworkers, married or single parents, were interviewed in their homes in the Coachella Valley. Two major problems identified were low wages and occupational exposure to pesticides. The average yearly income of the women interviewed was $5,000. Sixty percent of the women reported having pesticide related health problems. The results of this study do not support the assumption that the low socio-economic and political status of women farmworkers is due to the traditional gender value system of the Mexican American community. Rigid gender roles are not responsible for poverty and poor working conditions.

543. Lucas, Kimberley. *Differential Innovation Adoption Patterns of Female and Male Smallholder Farmers: A Case Study from the Taita Hills of Kenya.* M.A., The American University, 1986.

544. Machum, Susan T. *The Impact of Agribusiness on Women's Work in the Household, On-the-Farm and Off-the-Farm: A New Brunswick Case Study.* M.A., Dalhousie University, Canada, 1992.

A case study of women in potato farming in the Upper St. John River Valley in New Brunswick, Canada, discovered that women's work activities have indeed changed the structural transformation in agriculture and varied within time periods according to the kind of farm operations of which they were a part. Researchers need to pay greater attention to the differences among farm women and to recognize that women's off-farm work, on-farm work, and reproductive work requires a more complex framework than the urban dichotomy of productive (wage) and reproductive (domestic) work. Farm women's labor has been overlooked in the examination of the cost-price squeeze, producing a distorted picture of Canadian agriculture.

545. Manyeh, Marie Aliena. *Factors Constraining Women Farmers' Access to Agricultural Services in Sierra Leone.* M. Sc., University of Alberta, Canada, 1990.

Examines the factors constraining women farmers' access to extension services/information, training, and farm loans in Sierra Leone. The problems women faced in obtaining these agricultural services were found to be mainly socio-cultural, discriminatory, and extension/government related. Socio-cultural factors emerged as the greatest barrier to women's access to extension/information, while discrimination and extension/government related factors accounted for major problems of obtaining extension training and farm loans, respectively. The main suggestions made by the women to enhance their access to these services included a focus on themselves as a target group, the promotion of radio talks, monitoring extension activities, training more female extension workers, support and encouragement from husbands, and the promotion of group work among relatives, friends, and neighbors.

546. Margai, Magdalena T. *Factors Related to the Perception of Mastery of Alberta Farm Women.* M.Sc., University of Alberta, Canada, 1991.

The specific purpose of this study was to conduct a secondary analysis of data collected from Keating and Doherty's 1985 study of Alberta grain farmers in order to determine if demographic factors such as age, education, income, and farm land ownership influenced farm women's sense of mastery, focusing particularly on the sub-sample of 329 women who completed the third portion of the survey instrument.

The major finding is that education is the only demographic variable significantly associated with farm women's sense of mastery. Implications for this finding were discussed in terms of raising the level of education of rural farm women in order to raise their sense of mastery and, hence, to generate their own personal and socio-economic development.

547. Matsumoto, Valerie Jean. *The Cortez Colony: Family Farm and Community Among Japanese Americans, 1919-1982.* Ph.D., Stanford University, 1986.

Traces the history of the Cortez Colony, a Christian Japanese agricultural settlement, founded by Kyutaro Abiko, an immigrant businessman in 1919. Examines changes and continuity in institution and gender roles over the span of six decades and seeks to discover what factors enable this particular community to preserve its ethnic identity and unity from its inception to the present. Data were gathered from community institutional records and local newspapers to the World War II era Merced Assembly Center and Amache Relocation Camp publications and memorabilia. Eighty-one oral history interviews were conducted in 1982 with three generations of men and women. The study evidences significant changes over time including alterations in the relations between women and men and widening of career opportunities outside of the agricultural community.

548. Meiners, Jane Ellen. *Time Used for Household Work: Comparison for Farm, Rural Nonfarm, and Urban Women.* Ph.D., Oregon State University, 1984.

Examines how time is apportioned among various household work activities by farm women and compares time use by rural nonfarm and urban women. Data were from An Interstate Comparison of Urban/Rural Families' Time Use and consist of a stratified random sample of 2100 two-child families. Families were selected to equally represent five strata based on age of the youngest child. Time diary records were kept for each member of the family for a recall day and a record day. Farm women spent 49 hours per week on household work compared to 46 hours per week by urban women. Food preparation and dishwashing took the greatest portion of the time, about 15 hours per week. There was no significant difference in time spent on total household work. Farm women spent significantly more time on unpaid labor on the family farm or business than did rural nonfarm and urban women. Consideration was given to the implications for family decision making, public policy, and further research.

549. Mincolla, Joseph Anthony. *Early Career Development Processes of Women and Men Resource Managers in the USDA Forest Service.* Ph.D., Utah State University, 1988.

To understand the similarities and differences in the early careers of women and men resource managers of the United States Forest Service, it was hypothesized that differences would also be found in the early career goals of men and women in their ability to fit into an organization like the Forest Services and become contributing, productive members.

More similarities than differences were observed, but women had slightly different definitions of two important career goals, namely, service to an important cause and becoming competent managers. Both men and women possessed similar career goals and experienced similar levels of early career success. Immediate supervisors on first permanent Forest Service assignment had a much stronger influence on the early careers of the women in the study.

550. Morandy Nostrakis, Anny. *The Invisible Variable: Development Policy, Agriculture and the Role of Women in the Sahelian Region of Africa.* Ph.D., University of Delaware, 1991.

Examines the content of the neglect of the Sahelian women's contribution to subsistence agricultural production in West Africa and its impact on Sahelian development and identifies policy options for the future. Brings together and assesses the aggregated effects of the ideology and application of Neoclassical development theory, Sahelian governmental dynamics and agricultural policy, the sociocultural environment of women, and the policies and programs of international organizations that have responded to the food crisis in the Sahel. Women's conditions are used as a benchmark to measure the way in which agricultural policies undertaken in the Sahel have, by undermining women's economic status and perpetuating women's subordinate position, decreased food production and increased dependency and environmental deterioration. The exploitation of the rural sector by the Sahelian national government is detailed to demonstrate this. Concludes that agricultural self-reliance should be prioritized in order to overcome the current food crisis, and this cannot be reached without attending to the Sahelian women's role in agriculture and eliminating the gender inequities in the rural sector.

551. Olayiwole, Comfort Bang. *Rural Women's Participation in Agricultural Activities: Implications for Training Extension Home Economists.* Ph.D., Kansas State University, 1984.

Investigates to what degree rural women in Nigeria participate in agricultural production; how does this participation differ from one ethnic group to another; and to what degree do women farmers actually get the information on agricultural production available from agricultural extension services. Four villages, two predominantly Muslims and two non-Muslims, with a total of 120 households participated in the survey. The villages are all located in Kachia Local Government of Kaduna State. One adult female from each household was interviewed to determine household composition and farm labor by age and sex, male and female participation in each aspect of agricultural production, and access to agricultural land and supportive services, including agricultural inputs and information and the availability of home economic and technical occupational training from the extension services. Non-Muslim females of all age groups perform a majority of the farm tasks in all households of the two non-Muslim villages, while females participated in farm labor in only 12% in the Muslim village households. Women participate in the production of all crops as well as raising small livestock, particularly poultry and goats. Non-Muslim women also raise pigs.

It was found that despite women's participation in agricultural tasks and livestock raising, the agriculture extension agents do not address services to women. Home economics extension agents are not competent to teach women agricultural technical skills and are too few to reach the majority of rural households with the home economics information that women desire.

552. Parson, Karen. *Making Meaning, Making Butter: The Material World of Chester County Farm Women, 1750-1800*. M.A., University of Delaware, 1993.

Studies eighteenth century rural women's work and the material culture of that work, namely butter-making equipment. Analyzes the extent to which diversity existed in Chester County butter makers' experiences and broadly defines a multiplicity of meanings they have ascribed to their domestic tools. Evidence surrounding one woman, Elizabeth Smedley, her butter sales, and her butter-making equipment helps reveal the complexity of rural women's lives. Her butter making responded to family and farm changes; her equipment demanded maintenance and repair; and her customers varied over time. The analysis of this issue is based on public documents, prescriptive literature, tax records, craftsmens' and storekeepers' account books, and farm account books. Surviving eighteenth-century butter-making equipment provides crucial information about use, maintenance, and repair of domestic tools.

553. Phillips, Anne Radford. *Farm Women of Stokes County, North Carolina and the Production of Flue-Cured Tobacco, 1925 to 1995: Continuity and Change*. Ph.D., University of Maryland, 1960.

Explores the relationship of farm women in Stokes County, North Carolina, and looks at the relationship of farm women's lives and selected salient factors within the family, the county, and the larger national context. Analyzes women's roles within the family and in work patterns, the gender-linked power relationships within the family, and the values and reward systems of the farm women. Evidence points to a static culture in Stokes during the period 1925-1955, but the study explores how a number of large-scale factors energize the women, such as pattern of land ownership in the county and certain external forces such as the Depression and World War II.

Seeks to show the extent to which outside cultural forces did and did not lead to changes in family roles, power relationships, and values.

554. Pickett, Evelyn L. *Women in the Empty Quarter: A Study of Changes and Challenges as Related to Women's Experiences in the Nevada Ranching Industry from the Mid-Nineteenth Century to the Late Twentieth.* M.A., University of Nevada, Reno, 1988.

Examines the changes and challenges in the ranching industry in northeastern Nevada as they relate to women's experiences. Driven by a work ethic based on their own convictions, ranchwomen in the Empty Quarter, the rural areas of the state of Nevada, have taken the 8,000-year-old profession of livestock domestication and made the transition into a modern world. They span the gap between the old and the new.

555. Price, Lisa Marie Leimar. *Women's Wild Plant Food Entitlements in Thailand's Agricultural Northeast (Food Gathering Rights).* Ph.D., University of Oregon, 1993.

Examines wild plant food-gathering rights within the agricultural context and the central position of women in the formation and exercise of these rights. The data, generated from an in-depth village census and from participating in and observing village life, are based on field research from November 1989 to August 1990 among women from a wet-rice-farming village in Kalasin province in Northeast of Thailand.

Women's right to gather wild plant foods from agricultural lands depends upon concepts of private ownership associated with land tenure and a plant's assigned status. A plant species's status is based on taste desirability in association with collection purpose, species rarity, and market value.

Women farmers are engaged in the active protection, management, and propagation of wild-food plant species within their fields and around their settlements. Herbaceous plants, which are products of the new productive system that farm fields create, have the least restrictions relative to their numbers, while products from tree species have more gathering restrictions and are more frequently protected and transplanted within the farming environment. These rights and prohibitions are dynamic, collectively defined, and periodically re-evaluated by women. A woman's role as a gatherer, coupled with patterns of matrilocality and matrilineal inheritance of land, is a critical link to understanding this dynamic system of entitlement.

556. Quade, Ane Marie. *The Great Wheel and the Goose: Labor Transfer from Agriculture to the Textile Industry in the Early English Industrial Revolution.* Ph.D., University of Illinois at Urbana-Champaign, 1988.

Examines the impact of sex roles on early industrial development in England. The traditional sex-based division of labor was important in the growth of industry, but the economic relationship between men and women was changed as well. In pre-industrial England, men and women were expected to perform separate but complementary roles in the production process. Men were for the most part responsible for growing grain, while women produced dairy products.
As the enclosure movement expanded during the eighteenth century, the common pastures upon which the women's livestock industry depended were absorbed into large private holdings. Enclosures thus divorced women from their traditional sources of employment and income. Instead women turned to the employment of spinning for the textile factories. Data on wages in agriculture and the textile industry were gathered primarily from primary sources.

557. Reynolds, Lucile W. *Leisure-Time Activities of a Selected Group of Farm Women.* Ph.D., The University of Chicago, 1935.

558. Rome, Roman Albert. *Property and the Social Status of the Family Farmer in Gilded Age America.* Ph.D., City University of New York, 1984.

In nineteenth century America family farming was seen as the ideal calling of the American man and citizen, for it was widely assumed that farmers in the United Sates were almost universally the proprietors of the land they tilled, and the possession of a farm was believed to bring independence to its owner.
The census of 1880, the first to collect data on farm tenure, revealed that one farmer in four was not a proprietor, and it appears likely that by 1900 only half the farms in the United States were tilled by land-owning family farmers. The rest were worked by tenant farmers, sharecroppers, or landless

agricultural laborers. Investigation traces the impact of these developments on the social status of the family farmer in his local community and in the nation. Examines also the role which property played in defining relationships between the family farmer and his wife, his sons, and the hired man he sometimes employed.

559. Rose, Margaret Eleanor. *Women in the United Farm Workers: A Study of Chicana and Mexicana Participation in a Labor Union, 1950-1980.* Ph.D., University of California, Los Angeles, 1988.

Examines women's labor history and the interdependence of family, work, and unionism in women's lives. Previous studies on the United Farm Workers (UFW) have focused on male leadership and have provided a patriarchal interpretation of its origins, but this study documents the "invisible" contribution of women in this process. It shows how gender-defined forms of working-class activism shaped the involvement of Mexican-heritage women in UFW.

A wide variety and diversity of female participation exists in the UFW. Dolores Huerta, co-founder and first vice-president of the union, provides an example of a non-traditional female activist competing in the male-dominated sphere of decision-making. Huerta, however, is an exceptional case in the UFW. Most women in the union follow the example of Helen Chavez, wife of UFW president, Cesar Chavez. She demonstrates a traditional integration of family, work, and unionism. These two models of female unionism appear in various UFW operations. In the boycott the conventional pattern of female activism predominated. In the social services centers, women exerted authority but in the ranch committees women struggled for representation in a male-dominated sphere. Women's involvement in the most successful agricultural union in the United States changed their lives but did not fundamentally alter gender relations in the family, on the job, or in the union.

560. Sabol, Lois Malmberg. *A Role Analysis of Iowa Farm Families: Occupational Aspiration for Children, Job Satisfaction, and Women's Participation in Farm Work.* Ph.D., Iowa State University, 1986.

Investigation of the interface between work and the family may be particularly important on family farms. A system of mutually influencing roles composed of family, farm, and off-farm work roles filled by the members of the farm family was identified. This system was used as a perspective from which to study three aspects of contemporary farm life.

First, the parents' preference for a son to enter farming was investigated. Satisfaction with farming and the relative availability of resources to support a son's entry into agriculture were found to be predictive of a farming preference.

Second, satisfaction with farming was investigated, and a high degree of association was demonstrated among satisfaction levels with the farm and family role of the couple. The structure of satisfaction with farming differed depending upon the individual's participation in farm and off-farm work roles.

Third, the wife's participation in farm work was investigated. Results revealed that participation in other work roles was related to participation in the farm role.

561. Schipper-Peters, Caroline. *Agricultural Reforms in Lithuania: Implications for Rural Women, Farmers and Municipalities.* Ph.D., Iowa State University, 1993.

Directs attention to the roles and activities of various types of farmers, municipal officials, and rural women during the period of profound changes that are taking place in Lithuania.

The findings are presented in three different papers and explore first how the interests of regional and local municipal officials intersect with those of farmers. The findings indicate that local officials, due to lack of funds and know-how, are very limited in what they can do to assist farmers. The second paper focuses on gender and the effects of restructuring process on rural women's lives. The data gathered suggest that women are not playing an active role in the change process, and they feel they cannot spare the time to establish their own networks, even though they experience negative effects from the transition process. The third paper analyzes the changing relations between state and collective farms and new agricultural entrepreneurs and shows the linkages have a less competitive character than anticipated.

562. Seger, Karen Elizabeth. *Women and Change in the Yemen Arab Republic: A View from the Literature.* M.A., The University of Arizona, 1986.

563. Sharpless, Mary Rebecca. *Fertile Ground, Narrow Choices: Women on Cotton Farms of the Texas Blackland Prairie, 1900-1940.* Ph.D., Emory University, 1993.

Examines the contributions of women in the Texas Blackland Prairie, the most fertile southern region outside the Mississippi Delta. By 1900 this area was producing as much as one-sixth of the entire cotton production in the South. Immigrants from Germany, Czechoslovakia, and Mexico arrived in the early twentieth century joining the Anglo and African Americans who had come in large numbers after 1865. These women played critical roles in cotton farming. They contributed fiscally in two ways, first through the domestic economy, providing food, clothing, and other goods with their labor rather than purchase, and second, through the commercial economy as agricultural laborers in the cash crop. They also bore many children and produced the South's labor force. In many counties by 1920, tenant farmers or sharecroppers worked 75 percent of the land. Most women lived in extended families, and they spent much time trying to care for the children; many sewed all the family clothing; they also raised gardens, tended livestock, and cooked endlessly.

564. Smith, Deborah Beth. *Factors Influencing the Labor Force Participation and Labor Force Attachment Decisions of Minnesota Farm Women.* M.A., University of Minnesota, 1992.

Data from the Minnesota Farm Women Survey were used to investigate the contextual and perceptual factors which influence the labor force participation and labor force attachment decisions of 499 farm women. Management and feminist conceptual farmwork were the basis for the study. The findings indicated that labor force participation and labor force attachment were two distinct concepts influenced by different factors. The variables that were significantly related to labor force participation were not related to labor force attachment and vice versa. The farm women working off the farm full time did so for financial reasons, and the farm women who worked off the farm part time did so for personal reasons.

565. Smith, Rosslyn Braden. *Functions of Mass Media for Wisconsin Farm Women.* Ph.D., The University of Wisconsin, Madison, 1967.

566. Smuksta, Michael J. *Work and Family: Farm Women in Illinois 1820-1915.* Ph.D., Northern Illinois University, 1991.

Analyzes the work of farm women, gender relations within the household, family roles of farm women in commercial farm households in Northern Illinois, and security-first households in southern Illinois from 1820 to 1915, through diaries, family letters, and oral histories.

In frontier Illinois, farm women's household labor, household reciprocity, and production of petty commodities for local exchange satisfied immediate household needs or supplemented household income. This labor generated surplus income for the farm's entry into commercial agriculture. Increased market embeddedness in northern Illinois after 1840 shifted women's labor from household manufactures to the provision services associated with housework and childcare. Farm women became consumers. The need for household labor resulted in high fertility rates for frontier women during the antebellum period. The market served this link between production and reproduction in commercial farm households. Marriage became companionate; middle-class rural parents perceived their offspring as individuals; and couples limited family size.

These needs created inherent tensions within households founded upon a gender hierarchy and eroded the husbands' authority. Conflict characterized gender relations within commercial farm households of northern Illinois. Farm women expressed dissatisfaction over their workload, the allocation of limited resources for farm rather than home needs, and the husband's control over household income, especially that generated from petty commodity production. This capitalist transformation bypassed security-first households in southern Illinois. A productive strategy of relative independence from market relations depended upon women's household labor. High fertility rates persisted in these households. Farm women in southern Illinois had a poorer material standard of living than their counterparts in northern Illinois.

567. Staudt, Kathleen Ann. *Agricultural Policy, Political Power, and Women Farmers in Western Kenya.* Ph.D., The University of Wisconsin, 1976.

Looks at the political system as a mechanism for the exchange of resources and how it relates to women farmers. Little is known about the analysis of women's political activities, and women perform critical economic functions both within and outside the domestic sphere, but their work is understudied despite the significance of their activity. In much of Africa, women are the primary cultivators, managers, or co-managers of farm operations, and existent evidence points to the increased work responsibilities of these women with the advent of commercialized agriculture and male migration. Women often invest more time and labor in farm operations than men, and this means that the effectiveness of agricultural policy performance depends on reaching women farmers.

Despite these essential policy-related characteristics, women often receive considerably less access to agricultural services. Several studies merely note these inequities, but most simply ignore women in agriculture, assuming that men are heads of households and primary decision makers.

In this study, politics and women are analytically related in three different respects. First, women, like a number of economically differentiated gender groups, are crucially affected by policies which will also influence the form and mode of political response taken by groups. This perspective is particularly appropriate in analyzing what are described as relatively powerless groups. Second, analysis is directed at the implementation of a particular policy service, agriculture, patterns of differential service delivery, and the consequences of inequitable patterns for both productivity and the relative political resources of different groups. Finally this study examines how political power influences patterns of service delivery and, as such, draws material from both the organization and group theory literatures.

The form of women's political power has significant consequences for the type and quality of service received and is found to be crucially interrelated to the socio-economic context as well.

568. Stone, Margaret Priscilla. *Women, Work and Marriage: A Restudy of the Nigerian Kofyar*. Ph.D., The University of Arizona, 1988.

Research conducted among a group of farmers known as the Kofyar of central Nigeria provides a case study which runs counter to the general consensus that the transition from subsistence to market agriculture has hurt women's independent agricultural enterprises and incomes.

The Kofyar farming system as a whole is outlined, and the system of independent production is described within this context. The recent history of the Kofyar is described, including their migration into an agricultural frontier, the adoption of yams as the primary cash crop, and the evolution of a complex set of mechanisms for mobilizing labor. Women's important contributions to all types of labor are linked to their access to labor for their own independent production. Kofyar women's activities are seen as an essential part of Kofyar development. The system in general has become more prosperous and women as important contributors to that prosperity are also benefiting as individuals from these changes.

569. Szychter, Gwendolyn Mary. *Farm Women and Their Work in Delta, British Columbia, 1900-1939.* M.A., Simon Fraser University, Canada, 1992.

Explores the economic contribution of farm women in the Fraser Delta through interviews as well as personal diaries, local newspapers, and government documents. The women, a total of 24, whose lives were explored shared many common characteristics, but an effort was made to convey a sense of these women as individuals as well as members of a larger group.

Farm wives had an important economic role on their farms. All women performed a core group of activities of a domestic nature, most of which took place within the family. Over half of the women were involved in support work which generated a small income, and seven women did farm work to help the family financially--only one woman did so by preference. Women's work was essential in this agriculturally-based economy, but their importance did not depend on being directly involved in agricultural production.

570. Tima, Grace Ngemukong. *The Impact of Extension on Women in Two Rural Areas in Cameroon.* M.Sc., University of Guelph, Canada, 1991.

Analyzes the similarities and differences between two groups in Cameroon as they pertain to extension availability and applicability. Data gathered from eleven households and two farming groups in the North Province and seven households and three farming groups in the Western Province found that the impact of extension services to the women in both provinces was severely limited by several factors such as residential proximity of clientele to the location of extension services, religious and cultural traditions, educational attainment, lack of transportation, lack of credit, and the inclination of the extension service to make decisions for farmers. It is recommended that the local people should be involved in setting research goals instead of relying only on extension agents and researchers.

571. Timber, Priscilla Jean Tomich. *The Relation Between the Adoption of Modern Farm Practices and the Participation of Farm Women in Farm Tasks*. Ph.D., Utah State University, 1981.

Tests Parsons' theory that with industrialism women become less responsible for the instrumental role. This test was accomplished by determining whether farm women on more modern farms spent less time in the instrumental role activities than farm women on less modern farms, whether sex-role specialization characterized the participation of farm women and their husbands in the instrumental role activities on the farm, and if so whether sex-role specialization was more prevalent on more modern farms than on less modern farms. Finally it also looks at whether the results of any discoveries about sex-role specialization on the farm followed Parsons' ideas about the instrumental and expressive role patterns. Research instruments were sent to 450 randomly selected farm women in Dunn County, Wisconsin. Sixty-one percent returned their questionnaires. Parsons' idea that with industrialism women are less able to commit themselves to the instrumental role was not supported in this study. There was support for Parsons in that sex-role specialization, where the wife's specialty was to pay farm bills and keep records, was more prevalent on modern farms than on less modern farms. Farm women without young children, with less education, and with older husbands were discovered to be high participators in the farm tasks.

572. Tupas, Lourdes Pedregosa. *Rural Filipino Women's Participation in Postharvest Operations and General Farm Tasks with Implications for Agricultural Extension Programs for Women*. Ph.D., Kansas State University, 1983.

Determines differences between two groups of rural farm women and two groups of extension personnel regarding the participation of rural farm women in general farm tasks, the causes and extent of food losses, preference for media, groups and times suitable for training, as well as opinions about effective policies and practices on postharvest technology. Two hundred farm women were randomly selected from a list of farm wives from 20 villages in the province of Nueva Ecija in the Philippines and 68 extension personnel were subdivided into agriculture, home extension, and rural youth agents from the Ministry of Agriculture. The extension personnel were randomly selected from a list of personnel of the same province from which the farm wives came. An interview questionnaire was used to collect data from the rural farm women, and a survey questionnaire was administered by the researcher to the extension personnel. The findings revealed an increased involvement of rural women in general farm tasks, postharvest activities and a high need of training for these tasks. Found differences in perception among three groups of extension personnel about rural farm women and their participation in general farm tasks, postharvest operations, and the use of effective policies and practices.

573. Vangile, Titi. *The Rural Afforestation Project in Zimbabwe: Effectiveness and the Reality of Women's Participation.* M.N.R.M., The University of Manitoba, Canada, 1991.

Assesses the extent to which the Zimbabwe Rural Afforestation Project (RAP) Phase I has been successful in meeting its objectives of providing education, information, and extension services to promote tree planting and management by communal farmers. The study focuses on the woodlot approach, its constraints and opportunities, and also seeks to determine the role of women in the RAP as well as constraints or impediments to their participation.
In addition the technical support provided to women, that is, training and information was assessed to determine whether it was adequate to sustain the afforestation effort in the future.
Conclusions drawn from the literature search show that women, because of their objective views about conditions in the communal areas, are, in fact, the agents through which the afforestation efforts have to be channeled.

574. Wagner, Maryjo. *Farm, Families, and Reform: Women in the Farmers' Alliance and Populist Party.* Ph.D., University of Oregon, 1986.

Focuses on women in Kansas, Nebraska, and Colorado and examines biographical sketches, diaries, newspapers, letters, speeches, economic tracts, membership rolls, minute books, novels, and poetry to portray women's contribution to the organization, philosophy, and political platforms of the Farmers' Alliance and Populist Party. Many of the women combined busy private domestic roles as wives and mothers with public lives as organizers, public speakers, journalists, writers, and politicians. Often their writings and speeches showed the virtues of gentle, morally uplifting motherhood and other traditional female values. At the same time, defying convention, they left home for long periods of time to campaign for the new party, often emphasizing temperance and woman suffrage. Although Populist women did not win suffrage and temperance planks at national Populist conventions, they did acquire valuable political experience in the public sphere and form important networks with other women.

575. Wangari, Esther. *Effects of Land Registration on Small-Scale Farming in Kenya: The Case of Mbeere in Embu District.* Ph.D., New School for Social Research, 1990.

Tests the assumption that land registration in Kenya gives farmers incentive to improve their farms. Farmers with title deeds, it is argued are able to acquire credit and invest in their farms, which results in increased land productivity, employment and income. Land registration according to this view is crucial for agricultural development in Kenya.

A sample of 180 small-scale farming households, 90 registered and 90 unregistered was taken from Mbeere in the Embu District. Comparing data on the types of households established that land registration had no positive effects. While most registered farmers felt more secure, land ownership did not translate into higher productivity. Not all registered households had required title deeds, and not all of those who did had obtained credit. Only a few small-scale farmers benefitted; none of them were women, although women in both male- and female-headed households participated more in agricultural production. Credit was found to have no effect on production. In fact, the more credit was obtained, the less production was realized; credit was thus used for nonfarm activities. Concludes that land registration is significant in bringing about agricultural development, particularly in the semi-arid areas considered here. For registration programs to have positive effects would require better communication networks between farmers and policy makers, recognition of women's rights to land ownership, and better credit accessibility for women who are primary agricultural producers.

576. Whittaker, Wesley Lloyd. *Wife's Off-Farm Employment and Its Impact on Perceived Economic Welfare of Farm Households.* Ph.D., University of Illinois at Urbana-Champaign, 1991.

Explores the extent to which nonfarm labor force participation of the wife contributes to the economic welfare of the farm household, identifies the relative importance of a subset of variables most likely to affect the economic welfare of the farm household, and determines the relative importance of socioeconomic and social-psychological variables as groups in explaining perceived income adequacy.

Data come from the Illinois sample of the S-191 regional research project on the effect of nonfarm employment on family economic productivity and functioning. The sample included 240 respondents, where at least one spouse was involved in farming, and the dependent variable was perceived income adequacy, obtained from the Illinois portion of the instrument. Three models of perceived well-being were tested. A stepwise regression analysis was also used to isolate an important subset of variables. Model I explained 45% of the variation in perceived income adequacy. Five of the fourteen variables in the general model were statistically significant; this included total household income, net worth, satisfaction with farm income, and satisfaction with housing. Labor force variables contributed significantly to explain perceived economic welfare of farm households. However, the remuneration of the wives' off-farm employment does not necessarily improve the well-being of the farm household. Additional research is suggested.

577. Wilson-Larson, Laurie. *Farm Women's Experience of Work: A Look at Their Definition of Work and Factors Affecting Task Participation.* M. Sc., University of Alberta, Canada, 1991.

Eight farm women were interviewed for this study, and not all of the tasks performed on the farm by these women were perceived as work. Usually the tasks most directly connected to economic gain or the family "livelihood" were considered to be work. An interesting finding was that many of the women were involved in paid work which took place on the family farm. Five conceptual factors include the effects of the environment, farm structure, family structure, social/community influences, and individual variables. An interaction between the various contexts was found, with the environmental context and farm structural context generally having greater influence in decision making than the other contexts.

578. Zappi, Elda Gentili. *Mobilization of Women Workers in the Italian Rice Fields (1861-1915).* Ph.D., New York University, 1985.

Focuses on the women weeders in the rice fields of the Po valley in northern Italy and argues that despite adverse circumstances in a male-dominated society, these women weeders managed to organize and engage in collective action to improve their working conditions. Rice cultivation expanded in this area at an accelerated pace at the end of the eighteenth century, bringing about major economic and social changes, especially a class conflict between the capitalist farmers and the wage laborers, including women. In the prewar decade the workforce, which included some men and children, was 100,000 strong. Weeders residing in the rice belt comprised only about 60% of the workforce, the rest having to migrate from neighboring regions. For forty days, at the end of spring, the weeders gathered in the rice fields to do this strenuous job. In addition to chronic low wages, hard work, and threat of unemployment, the weeders also suffered from health hazards on the job and competition from migratory labor. The Socialist organizers formed weeders' leagues, which, with various tactics, fought for workers' rights. In this struggle the weeders acquired a class consciousness, a political awareness, and a sense of solidarity which made them strive for higher wages, improved working conditions, and social justice.

RESEARCH GUIDE

FRENCH BOOKS AND ARTICLES

579. Als, G. "Les tendances d'évolution de l'agriculture et le rôle de la femme en agriculture" (Evolutionary Trends in Agriculture and the Role of Women in Agriculture), *Bulletin du Statec* 31:5 (1985):123-138.

    Reviews the problems of women in the evolution of farming structure, mechanization, and the agricultural workforce.

580. Barthez, Alice. "Femmes dans l'agriculture et travail familial" (Women in Agriculture and Domestic Work), *Sociologie du Travail* 26:3 (July-September 1984):255-267.

    Women's participation in agricultural labor in industrialized societies is analyzed, focusing on the family as the production unit. French census data from 1955 to the present are reviewed.

581. Barberis, Corrado. "La Femme dans l'agriculture italienne" (Women in Italian Agriculture), *Etudes Rurales* 10 (July-September 1963): 50-67.

    Examines women's role in Italian agriculture.

582. Becouarn, M.C. "Agricultrice: la maîtrise d'un métier" (Women Farmers: Mastery of a Career), *Cultivar* 200 (October 1986): 260-261.

    Describes women farmers' role in France.

583. Bergen, A., et al. "Les opportunités et les contraintes pour les produits cultivés aux périmètres irrigués et aux jardins potagers dans la vallée du fleuve Sénégal de la perspective du marketing" (Opportunities and Limitations for Agricultural Production from Irrigation Schemes and in Kitchen Gardens in the Sénégal River Valley from a Marketing Perspective), in *Design for Sustainable Farm-Managed Irrigation Schemes*

in *Sub-Saharan Africa*, v. 2, Ubels, J. ed. Wageningen, Netherlands: Wageningen Agricultural University, 1990.

Based on two studies, one on rice irrigation schemes and one on kitchen gardens, the authors examine the extent to which an approach with a marketing perspective can contribute to the development both of viable design for small-scale irrigated agriculture and of the adaptation of existing irrigation schemes.

584. Bisilliat, Jeanne, et al. *Femmes et politiques alimentaires: actes du séminaire international sur la place des femmes dans l'autosuffisance et les stratégies alimentaires* (Women and Food Politics), Paris: Edition d'ORSTOM, Institut Français de Recherche Scientifique pour le Développement en Coopération, 1985.

Discusses women's role in farm organization and management.

585. Cohen, Yolande. "Les métiers féminins: une conquête? Histoire comparée des fermières et infirmières dans l'entre-deux-guerres" (Women's Careers: A Conquest? A Comparison between Farmers and Nurses Between the Two World Wars), *Resources for Feminist Research* 15:4 (1986-87): 54-55.

Outlines an ongoing research project by the author.

586. Cormier, D. "Les syndicats de gestion agricole du Québec: un regroupement d'agriculteurs et agricultrices" (Agricultural Management Association in Quebec: A Regrouping of Farmers and Agricultural Women), *Canadian Journal of Agricultural Economics; Revue Canadienne d'Economie Rurale* 37:4 pt.1 (December 1989): 597-608.

Describes the farmers' associations' agricultural policy and examines the organizations' development in regard to extension activities, rural women, program development, and membership.

587. Corrèze, A. "Femmes paysannes d'Afrique: une rencontré" (An Encounter with Farm Women in Africa), *ORSTOM, Laboratoire de Sociologie et Géographie Africaines* 3 (1985): 97-103.

   Examines rural women's status in Africa.

588. Cros, M.-O. "Les besoins des agricultrices en matière de vulgarisation dans une zone de grande culture: le G.V.A.M. du Plateau d'Evreux-St-André" (Farmers' Wives Advisory Needs in a Large Farming Area: The GVAM of Evreux-St-Andre), *Economie Rural* 95 (1973): 37-41.

   A case study examines how the agricultural extension services have failed in the case of women.

589. Culaud, H.P. "Les modification du régime" (Changing the Regime), *Paysans* 184 (1987): 49-51.

   The agricultural reforms under discussion in France envisage that social security provisions for female farmers will be brought into line with those for males, and farm women's position will be protected along with widows' pensions.

590. Dentzer, M.T. "Formation des agricultrices: un montage audio-visuel dans les Vosges" (Training of Women Farmers: An Audio-Visual Assemblage in Vosges), *Inform Ari* (Paris) 435 (April 1973): 50.

   Describes audiovisual training of women farmers.

591. Donner-Shabafrouz, I. "Le rôle des femmes dans les systèmes traditionnels d'agriculture et leur intégration aux projets de développement de l'agriculture irriguée" (The Role of Women in Traditional Agriculture and Their

Integration in Modern Irrigation Projects), *Bulletin: Deutscher Verband fur Wasserwirtschaft und Kulturebau* 8 (1983): 139-154.

Women's role in traditional farming and their integration into irrigation projects, including technical training.

592. Everaet, H. *La situation sociale de la fermière: sa participation au travail de l'exploitation* (Social Situation of the Farmer's Wife: Her Participation in the Work of the Farm), Brussels: Ministere de l'Agiculture, Cahiers de l'Institut Economique Agricole, No 139/RR-117, 1972.

Structural improvements in the running of both family and farm which would be both labor saving and lighten the work load are suggested. Modern management provides an opportunity for the farmer's wife to become more involved in the administrative aspects and less in the physical duties of operating the farm.

593. Everaet, H. *Les agriculteurs et agricultrices après la reprise de l'exploitation* (The Effect of Taking up Farm Ownership on Men and Women in Farming), Belgium: Cahiers de l'Institut Economique Agricole, Ministere de l'Agriculture, 1985.

Looks at the demographic and economic characteristics of newly established farm owners in Belgium and the basis of statistical data collected on farmers who attained ownership between 1978-1981.

594. Ferchiou, Sophie. *Les femmes dans l'agriculture tunisienne* (Women in Tunisian Agriculture), Aix-en-Provence: Edisud, 1985.

Women agricultural laborers in Tunisia and rural conditions.

595. Filippi, G., and Nicourt, C. "Cohérence et professionnalité dans le travail des agricultrices d'une commune de Dordogne" (Consistency and Professionalism

in the Work of Women Farmers in a Dordogne Commune), *Actes et Communications, Economie et Sociologie Rurales, Institut National de la Recherche Agronomique* No.3 (1988): 85-98.

Demonstrates the capabilities of women and the work they achieve in a village of Perigord. Examines how the women organized their own work in order to contribute to the overall production of their farms.

596. Geschière, P. "L'agriculture de subsistance, l'autonomie de la femme et l'autorité des aînés chez les Maka (Cameroun)" (Subsistence Agriculture, Women's Autonomy and the Authority of the Elders of the Maka Tribe (Cameroon)), 3 *Journal d'Agriculture Traditionnelle et de Botanique Appliquée,*0:3-4 (1983): 307-320.

Describes food production and women's status among the Maka people of eastern Cameroon.

597. Granie, A.M., and P. Roux. "Les agricultrices en question" (Focus on Women Farmers), *Nouvelles Campagnes* 35 (1985): 13-23.

Investigation of women farmers in France and their view of themselves and the significance of the involvement of women in farm accounting is also discussed.

598. Guillou, Anne. *Les femmes, la terre, l'argent* (Women, Earth, Money), Brasparts: Editions Beltan, 1990.

Abridged version of the author's thesis (Doctorat d'Etat, Université de Nantes, 1987). Economic and social conditions of women working in agriculture.

599. Helsloot, L. "La contribution de la conception à l'autogestion par les femmes: l'exemple des jardins potagers des groupements de femmes sur l'Ile a Morphil au Sénégal" (Contribution of Women to the Idea of Self-Management: The Kitchen Gardens of Women's Groups on Senegal's Ile a Morphil) in *Design for Sustainable Farm-Managed Irrigation Schemes in Sub-Saharan Africa*, v. 1., Ubels, J., ed., Wageningen, Netherlands: Wageningen Agricultural University, 1990.

   The introduction of a women's irrigated kitchen garden project on Senegal's Ile a Morphil has resulted in improved self-confidence, growth of women's organizational capabilities, and an improvement in women's roles.

600. Herzog, W. and H. Muhl. "Eléments indispensables à une bonne formation de l'agricultrice et de l'agriculteur" (Vital Elements in Training Men and Women in Agriculture), *Publication de la Confédération Européenne de l'Agriculture, CEA* 61 (1979): 47-83.

   Reflects the CEA's efforts to improve vocational training for men and women in agriculture.

601. Lagrave, Rose-Marie. "Bilan critique des recherches sur les agricultrices en France," (Critical Assessment of Research on French Women Farmers), *Etudes Rurales* 92 (October-December 1983): 9-40.

   Examines multidisciplinary research on French women farmers.

602. Lagrave, Rose-Marie, ed. *Celles de la terre. l'agricultrice: l'invention politique d'un métier* (Women of the Land. The Female Farmer: The Political Invention of an Occupation), Paris: Editions de l'Ecole des Hautes Etudes en Sciences Sociales.

   This book was reviewed in *Cahiers Internationaux de Sociologie* 36:86 (January-June 1989):183-184.

603. Lasram M., and P. Plaza, eds. *La vulgarisation, composante du développement agricole et rural* (Extension as a Component of Agricultural and Rural Development), Montpellier: Centre International des Hautes Etudes Méditerranéennes, Cahiers Options Méditerranéennes 2:4, 1994.

   Reviews progress in extensions in relation to agricultural development in the Mediterranean countries, formulates policies, and identifies possible areas for collaboration in the future.

604. Leplaideur, S. *"Le long voyage du karite: shea nuts."* Inter Tropique 21 (1987): 21-23.

   Argues that the state and processing firms in Europe make exhorbitant profits on the shea nuts used in the production of highly expensive cosmetic products and harvested by African women. In Burkina Faso, where shea nuts constitute the country's third export product, the state-controlled marketing system provides no financial incentives to these women.

605. Lindemann-Meyer zu Rahden, H. "Collaboration entre paysan et paysanne au sein d'une exploitation moderne" (Collaboration Between Men and Women Farmers on a Modern Farm), *Confed Eur Agr Publ* 45 (1972):186-191.

606. Maresca, Sylvain. *L'autoportrait de six agricultrices en enquête d'image* (The Self Portrait), Presses Universitaires du Mirail; Paris: Institut National de la Recherches Agronomique, 1991.

   Portraits of six women farmers.

607. Martin S. "Le statut de l'agricultrice aujourd'hui conditionne l'avenir de l'agriculture" (The Status of Women Farmers Today Conditions the Future of Agriculture), *Paysans* 19:113 (August/September 1975):35-41.

   Reports on the current status of women in French agriculture.

608. Nicourt, Christian, and Geneviève Filippi. "Contribution à la définition d'un métier: agricultrice" (Contribution to the Definition of an Occupation: Woman Farmer), *Sociologie du Travail* 29:49 (1987):477-494.

   The differences between prescribed and actual work of farm wives are analyzed by observation.

609. Nicourt, Christian. "Contribution à l'étude du temps de travail cohérence et durée dans le travail des agricultrices" (A Study of the Consistency and Length of Working Time of Women Farmers), *Economie Rurale* 210 (1992): 44-56.

   Discusses women farmers' working time allocation with specific reference to French farm women.

610. Nicourt, Christian, and Olivier Souron. *Temps et rythmes des cultivatrices: le travail des femmes à Marcillac-Saint-Quentin en Périgord 1900-1939* (Time and Rhythms of Female Farmers), Paris: INRA, Station d'Economie et Sociologie Rurales; Centre de Recherches Cadre de Vie-Cadre de Travail, 1988.

   Historical studies of French women farmers from the 20th century.

611. ORSTOM, Laboratoire de sociologie et géographie africaines. *Dynamique des Systèmes Agraires* (Dynamics of Agrarian Systems), Paris: Editions de l'Orstom, 1985.

   Covers women and technology in rural development.

612. Rattin, S. "Vers une féminisation de la fonction de chef d'exploitation" (An Increasing Proportion of Women Are Farmers), *Cahiers de Statistique Agricole* 5/6 (1985): 33-45.

   Examines the increase of women in farm management position in

France since 1975, reaching 11% in 1983. Some of these women are old widows who have taken over the farm after their husbands died, and others are young married women whose husbands are engaged in non-farming activities.

613. Rialland-Morissette, Yvonne and Ghislaine Desjardins. *Le passé conjugué au présent* (The Past Joined the Present), Montréal: Pénélope, 1980.

    Describes women farmers in Quebec from 1915 to 1980.

614. Roberts, P. "Les femmes et les programmes de développement rural avec référence aux programmes-femmes financés par le Fonds Européen de Développement au Kenya" (Women and Rural Dévelopment Programmes Financed by the European Development Fund in Kenya), *Revue Tiers Monde* 26:102(1984):299-305.

    Argues that rural development programs in Africa tend to overlook the interest of female farmers and may even operate to the detriment of such interests.

615. Rossier, R. "Attitude de la paysanne face à la mécanisation de l'exploitation" (The Attitude of the Female Farmer Towards Mechanization on the Farm), *Technique Agricole* 53:15 (1991): 14-20.

    Examines mechanization in milking, waste disposal, forage feeding, and hay unloading on 622 farms in different regions of Switzerland and their effects on women farmers. Their attitudes towards mechnization are also investigated.

616. Schwartz, A. "Modernisation de l'agriculture et transformation des rapports sociaux de production chez les Ngam-Ngam et les Tchokossi du Nord-Togo: l'impact du 'Projet Namiele'" (Agricultural Modernization and Transformation of Social Relations of Production among the Ngam-Ngam and Tchokossi in Northern Togo: Impact of the Namiele Project), *Cahiers des Sciences Humaines* 25:3 (1989): 357-367.

The Namiele project's activities are briefly presented. It encouraged rather than forced farmers to adopt an individualistic approach to their role within the production system, with a view of promoting modern farming systems.

617. Stevens, Hélène, and Anne-Marie Jeay. "Femmes d'Afrique et des pays méditerranéens: le travail agricole des femmes dans la région pauvre des pays méditérranéens et africains" (Women in the Poor Region of Mediterranean and African Countries), *Sociologia Ruralis* 18:4 (1978): 235-244.

Examines agricultural work done by women in the Mediterranean areas of Europe and Africa.

BIBLIOGRAPHIES ON WOMEN IN AGRICULTURE

618. Ardayfio-Schandorf, Elizabeth, and Kate Kwafo-Akoto. *Women in Ghana: An Annotated Bibliography*. Accra New Town: Woeli Publishing Services, 1990.

   Has one chapter on Agriculture and Environment. All citations are theses for degrees presented at the University of Ghana, and they are all very well annotated. The document provides a guide to scholars, policy makers, and development agents interested in integrating the female population in development. The major aim of the study is to provide information on the labor force participation of Ghanaian women. The study also determines areas on women not yet covered by research.

619. Ashby, Jacqueline A., and Stella Gomez. *Women, Agriculture, and Rural Development in Latin America*. Colombia: Centro Internacional de Agricultura Tropical, 1985.

   Includes a partly annotated bibliography with 415 entries from Latin America covering studies in Spanish and English.

620. Baas, Ettie. *Women in Development: A Bibliography*. The Hague, Netherlands: Institute of Social Studies, fifth revised edition, 1980.

621. Bindocci, Cynthia Gay. *Women and Technology: An Annotated Bibliography*. New York: Garland Publishing, 1993, pp. 42-50.

   Includes twenty-seven entries related to "Agriculture and Food Technology."

622. Bullwinkle, Davis A. *African Women: A General Bibliography, 1976-1985*. Westport, CT: Greenwood Press, 1989.

Organized under large subject headings, this bibliography presents citations to English-language publications about women in Africa during the United Nations Decade for Women from 1976-1985. Several chapters have information on Women in Agriculture; however, many citations are listed in several places. No annotations are included.

623. Buvinic, Mayra, et al. *Women and World Development: An Annotated Bibliography.* Washington, D.C.: Overseas Development Council, 1976.

   The concepts central to most of the works in this bibliography are "status of women" and "women's role." Chapter four, "Socio-Economic Participation of Rural Women" and chapter six, "Women's Work and Economic Development" are specifically related to agriculture. All entries are very well annotated, and the material covers books, journal articles, and conference papers. Each chapter is organized under large geographical headings.

624. Cebotarev, E. A., et al. "An Annotated Bibliography on Women in Agriculture and Rural Societies." *Resources for Feminist Research* 11 (March 1982): 93-180.

   Includes published and unpublished documents from both academic and non-academic sources. Organized geographically and covers studies in five languages although English predominates.

625. Center for Policy and Development Studies. *An Annotated Bibliography of Women in Farming Systems in the Philippines.* College, Laguna: Center for Policy and Development Studies, U.P. at Los Banos in Cooperation with the International Rice Research Institute, 1988.

626. Cheng, Lucie, et al. *Women in China: Bibliography of Available English Language Materials.* Berkeley, CA: Institute of East Asian Studies, University of California, Center for Chinese Studies, 1984.

A list of 4,100 citations under large subject headings. There are over 130 citations on "Rural and Agricultural Labor." The entries are not annotated.

627. Craig, R.A. *Rural Women: An Annotated Bibliography of Australian Sources.* Australia: Department of Extension and Education, Roseworthy Agricultural College, 1980.

Covers the changing position of farm women, the contribution of education and employment to changing roles for women in rural areas, the extent of women's participation in rural organizations, and the influence of women in rural communities.

628. Danforth, Sandra. "Women, Development, and Public Policy: A Selected Bibliography" in *Women in Developing Countries: A Policy Focus*, eds., Staudt, Kathleen, and Jane S. Jaquette. New York: The Haworth Press, 1983. Also published in *Women & Politics* 2 (Winter 1982):107-124.

Includes published academic literature in English since 1960 on women in all regions of the contemporary developing world. An interest in public policy and politics has influenced the selection. The entries are not annotated.

629. Fairbanks, Carol, and Sara Brooks Sundberg. *Farm Women on the Prairie Frontier: A Sourcebook for Canada and the United States.* Metuchen N.J.: Scarecrow Press, 1983.

Four essays with many references introduce the readers to the land and the people, the history of the region, and the fiction. The references related to "A Usable Past: Pioneer Women on the American Prairies," "Farm Women on the Canadian Prairie Frontier: The Helpmate Image," and the annotated list of references on "History and Background" are specifically related to women in agriculture.

630. Faris, Mohamed A., and Mahmood Hasan Khan, eds. *Egyptian Women in Agricultural Development: An Annotated Bibliography.* Boulder: Lynne Rienner Publishers, 1994.

There is increasing recognition in Egypt of the need to integrate women, particularly in the rural areas. In this bibliography we find 175 annotated documents including 53 books, 75 journal articles, 10 graduate theses, 24 technical reports, and 13 papers. Some of the citations cover multiple issues and refer to Africa in general.

631. Fera, Darla. *Women in American Agriculture: A Selected Bibliography.* Washington, D.C.: U.S. Department of Agriculture, 1977.

Covers studies of women's activities on the farm, specifically, women engaged as landowners, farm managers, agricultural laborers, and in agriculture industries, such as bee-keeping, silk culture, butter production, etc. The entries are not annotated.

632. Fortmann, Louise. *Tillers of the Soil and Keepers of the Hearth: A Bibliographic Guide to Women and Rural Development.* Bibliographic Series, no. 2. Ithaca, N.Y.: Rural Development Committee, Center for International Studies, Cornell University, 1979.

Includes information from the Third World as well as the United States. Prepared for the teaching faculty giving courses on women and rural development, undertaking research, or designing programs in the field. Includes citations to books, articles, and a number of unpublished papers and international documents.

633. Hafkin, Nancy J. *Women and Development in Africa: An Annotated Bibliography.* Addis Ababa: United Nations, Economic Commission for Africa, 1977.

Organized by country and author index, this bibliography includes sections on general studies on women and development in Africa, rural development and women, population studies, education and training, urban development and women, and women's organizations.

634. Hombergh, Helen van den. *Gender, Environment and Development: A Guide to the Literature*. Amsterdam: International Books, 1993.

Connections between gender, environment, and development (GED) have developed since the late 1980s, and Hombergh looks at gender as it refers to cultural and historical concepts of femininity and masculinity and the power relations between men and women. The term environment refers to natural resources with emphasis on their close relation with the macroeconomic, political, and cultural environment, and the term development refers to the transition from poverty to wealth (economic development). The book is divided into two parts, introductory texts on various issues related to GED and an extensive bibliography. A list of journals related to the topic and a subject index are also included.

635. Jiggins, Janice. "Agricultural Technology: Impact, Issues, and Actions" in Gallin, Rita S., et al., ed. *The Women and International Development Annual* v.1. Boulder: Westview Press, 1989, pp. 25-55.

Reviews the relationship between women and the technological changes that are taking place and gives directions for future research and policy.

636. Joint Bank-Fund Library. *Women and Development: Articles, Books and Research Papers Indexed in the Joint World Bank-International Monetary Fund Library, Washington, D.C.* Boston: G.K. Hall & Co. 1987.

Includes international material from 1977 to 1986. Some entries are under the heading of agriculture. Indexed by author, title, and subject.

637. Joyce, Lynda, and Samuel M. Leadley. *An Assessment of Research Needs of Women in the Rural United States: Literature Review and Annotated*

Bibliography. University Park: Pennsylvania Agricultural Experiment Station, Pennsylvania State University, 1977.

Discusses methods and theories for research on women in rural areas of the United States from 1900 to 1975. Includes a 30-page annotated bibliography covering popular literature, monographs, and agricultural bulletins as well as research studies.

638. Kessler, Shelly. *Third World Women in Agriculture: An Annotated Bibliography*. New York: National Council for Research on Women, 1985.

Covers fugitive literature prepared from conferences and in-house studies and gives a list of collections accessed. All material is located in special libraries but is available to the general public. The purpose of this literature search was to locate the various materials on this subject available in the New York-Washington area.

639. King, Evelyn. *Women on the Cattle Trail and in the Roundup*. Bryan, TX: Brazo Corral for the Westerners, 1983.

Short essay on women's involvement on the cattle trail and in the roundup and a detailed bibliography with short annotations on the same subject.

640. Kinnear, Mary, and Vera Fast. *Planting the Garden: An Annotated Archival Bibliography of the History of Women in Manitoba*. Winnipeg, Manitoba: The University of Manitoba Press, 1987.

A few entries on women in agriculture on pages 122-124.

641. Lewis, Martha Wells. *Women and Food: An Annotated Bibliography on Family Food Production, Preservation, and Improved Nutrition*. Washington D.C.: Office of Women in Development, U.S. Agency for International Development, 1981.

Includes studies on agricultural extension work, sexual division of labor, and nutrition in developing countries.

642. Mascarenhas, Ophelia, and Marjorie Mbilinyi. *Women in Tanzania: An Analytical Bibliography.* New York: Africana Publishing Co., division of Holmes & Meier, 1983.

Includes a chapter analyzing the struggles of women peasants, sexual division of labor within the fields of agriculture, and the struggle taking place over allocation of land. The references cover citations to unpublished papers, master's theses, journal articles, chapters in books. All citations are annotated.

643. Nelson, Nici. *Why Has Development Neglected Rural Women? A Review of the South Asian Literature.* Oxford: Pergamon Press, 1979.

A review of the literature available on the role of women in rural development in South Asia. Starts with a definition of what the author understands development to be and the reasons for considering the role of women in development. Assesses in detail how much we presently know about rural women and why, how, and with what impact women do or do not participate in development. Examines the quality and quantity of research material available.

Suggests a selected list of possible avenues for future research.

644. Nwanosike, Eugene O. *Third World Women and Rural Development: A Selected Bibliography.* Doula, Cameroon: Pan African Institute for Development, 1984.

Covers rural development studies on women in the Third World.

645. Pescatello, Ann. "The Female in Ibero-America: An Essay on Research Bibliography and Research Directions." *Latin American Research Review* 7 (Summer 1972):125-141.

Gives overview of women's place in Latin American society and how the Ibero-American scene is colored by its own peculiar mix of cultural antecedents and traditional perceptions. Reviews new studies on women in Ibero-America and includes a bibliography organized by country. Citations are both in English and Spanish, most of them for books, but they are not complete since they lack publishers' names.

646. Rafats, Jerry. *Women in Agriculture.* Beltesville, MD: U.S. Dept. of Agriculture, National Agricultural Library, 1989.

A listing of 361 citations entered in AGRICOLA, the database of the National Agricultural Library, United States Department of Agriculture from 1979 to March 1989. Most of the information has a short annotation. All materials listed are available from the National Agricultural Library through Inter Library Loan.

647. Rihani, May. *Development as If Women Mattered: An Annotated Bibliography with a Third World Focus.* Washington, D.C.: Overseas Development Council, 1978.

Includes 287 entries most for English-language items including information published in the 1970s, arranged into subject categories and subdivided geographically.

648. Tinker, Irene, et al. *Women and World Development. With an Annotated Bibliography.* New York: Praeger Publishers, 1975.

The second part of this book gives "A Critical Review of Some Research Concepts and Concerns" by Mayra Buvinic and includes a large bibliography on all topics on women in development. Some are related to rural women and agriculture.

649. University of West Indies, Barbados. *Planning for Women in Rural Development: A Source Book for the Caribbean, 1976-1985.* Barbados: University of West Indies, 1985. Includes studies on rural women in the Caribbean.

650. Vavrus, Linda Gire, and Ron Cadieux. *Women in Development: A Selected Annotated Bibliography and Resource Guide*. East Lansing, MI: Institute for International Studies in Education, College of Education, Michigan State University, 1980.

>   Includes a few studies on agriculture and food production.

651. Weintraub, Irwin. *Black Agriculturists in the United States 1865-1973: An Annotated Bibliography*. University Park, PA: The Pennsylvania State University Libraries, 1976.

>   Includes a few studies related to women.

652. Whitehead, Vivian B. *Women in American Farming: A List of References*. Davis, CA: Agricultural History Center, University of California, Davis, 1987.

>   Contains citations to over 1000 periodical articles, books, and government publications on women in American agriculture. The author gathered the information from works listed in the National Agricultural Library, Library of Congress, and the Agricultural History Branch of the USDA. The book is arranged alphabetically by author with a brief annotation accompanying most citations.

LIST OF JOURNAL TITLES FREQUENTLY PUBLISHING ON WOMEN IN AGRICULTURE

We have included a list of journals which frequently publish articles on the topic of women farmers or women working in the fields of agriculture. We thought this might be of interest to researchers who would like to subscribe to a journal and did not have a library nearby to look up the address, etc. We have included the year the journal started publishing, and the number of issues per year, address, telephone and fax numbers when available as well as some of the indexes where they are indexed. We are not suggesting that these are the best journals; their inclusion is based only on the number of articles listed on our topic.

653. *Agricultural Economics.* v.1 1986, 6/year
      Elsvier Science B.V.
      P.O. Box 211
      1000 AE Amsterdam, The Netherlands
      Tel: 31-20-5803911
      Fax: 31-20-5803598
      Indexed in: *World Agricultural Economic & Rural Sociology Abstracts; Maize Abstracts, Tropical Oil Seeds Abstracts.*

654. *Agricultural Education Magazine.* v.1 1929, 12/year
      Agricultural Education Magazine Inc.
      2441 Suzanne Drive
      Mechanicville, VA 23111-4028
      Tel: 804-746-3538
      Indexed in: *CIJE; Current Contents; Education Index; Farm and Garden Index.*

655. *Agricultural History.* v.1 1927, 4/year
      Agricultural History Society
      University of California Press, Journal Division
      2120 Berkeley Way, Berkeley, CA 94720
      Tel: 510-643-7154
      Fax: 510-642-9917
      Indexed in: *America: History & Life; Biological & Agricultural Index; Environmental Abstracts; Historical Abstracts; Dairy Science Abstracts.*

656. *American Anthropologist.* v.1 1899, 4/year
American Anthropological Association
4350 N. Fairfax Drive, Ste. 640
Arlington, VA 22203
Tel: 703-528-1982
Indexed in: *Abstracts in Anthropology; America: History & Life; Biological & Agricultural Index; Historical Abstracts; Social Science Index; Studies of Women Abstracts.*

657. *American Journal of Agricultural Economics.* v.1 1919, 5/year
American Agricultural Economic Association
80 Heady Hall
Iowa State University, Ames, AI 50011-1070
Tel: 515-294-8700
Fax: 515-294-1234
Indexed in: *Biological & Agricultural Index; Current Contents; Nutrition Abstracts; World Agricultural Economic & Rural Sociology Abstracts; Sociological Abstracts.*

658. *Anthropological Quarterly.* v.1 1928, 4/year
Catholic University of America Press
620 Michigan Avenue N.E.
Washington, D.C. 20064
Tel: 202-319-5052
Fax: 202-319-5802
Indexed in: *Anthropological Literature, Current Contents; Rural Development Abstracts; Social Science Index; World Agricultural Economic & Rural Sociology Abstracts; Studies on Women Abstracts.*

659. *Canadian Journal of African Studies.* v.1 1967, 3/year
Canadian Association of African Studies
Center for Urban and Community Studies
University of Toronto
455 Spadina Avenue, Ste. 426
Toronto, Ontario, M5S 2G8
Canada
Tel: 613-237-6885
Fax: 613-237-2105

Indexed in: *Abstracts in Anthropology; America: History & Life; Sociological Abstracts; Historical Abstracts; Social Science Citation Index; World Agricultural Economic & Rural Sociology Abstracts.*

660. *Current Anthropology.* v.1 1966, 5/year
Wenner-Gren Foundation for Anthropological Research
University of Chicago Press, Journal Division
5720 S. Woodlawn Avenue
Chicago, IL 60637
Tel: 312-753-3347
Fax: 312-753-0811
Indexed in: *Abstracts in Anthropology; America: History & Life; Biological Abstracts; Historical Abstracts; Modern Language Association Abstracts; Social Science Citation Index; Studies on Women Abstracts.*

661. *Development and Change.* v.1 1969, 4/year
(Institute of Social Studies, The Hague, NE)
Blackwell Scientific Publication, Ltd.
Osney Mead, Oxford OX2 0EL
England
Tel: 0865-240201
Fax: 0865-721205
Indexed in: *America: History & Life; Current Contents; Historical Abstracts; Rural Development Abstracts; Social Science Citation Index.*

662. *Economic Development and Cultural Change.* v.1 1952, 4/year
University of Chicago Press, Journal Division
5720 Woodlawn Avenue
Chicago, IL 60637
Tel: 312-753-3347
Fax: 312-753-0811
Indexed in: *Abstracts in Anthropology; America: History & Life; Historical Abstracts; Rural Development Abstracts; Urban Studies Abstracts; Sociological Abstracts; World Agricultural Economic & Rural Sociology Abstracts.*

663. *Ethnology: An International Journal of Culture and Social Anthropology.* v.1 1962, 4/year
University of Pittsburgh
Department of Anthropology
Pittsburgh, PA 15260
Tel: 412-648-7593
Fax: 412-648-5911
Indexed in: *America: History & Life; Anthropological Literature; Current Contents; Historical Abstracts; Social Science Index; World Agricultural Economic & Rural Sociology Abstracts.*

664. *Human Organization.* v.1 1941, 4/year
Society of Applied Anthropology
Box 24083
Oklahoma City, OK 73124-0084
Tel: 405-843-5113
Indexed in: *Abstracts in Anthropology. America: History & Life; Current Contents; Historical Abstracts; Rural Development Abstracts; World Agricultural Economic & Rural Sociology Abstracts.*

665. *Journal of African History.* v.1 1960, 3/year
(Text in English and French)
Cambridge University Press, Edinburgh Bldg. Shaftesbury Rd.
Cambridge, CB2 2RU
England
Tel: 0223-312393
Fax: 0223-315052
North American Address:
Cambridge University Press, Journals Dept.
40 W. 20th Street
New York, N.Y. 10011
Tel: 212-924-3900
Fax: 212-691-3239
Indexed in: *Anthropological Literature; America: History & Life; Current Contents; Historical Abstracts; Humanities Index; Social Science Citation Index.*

666. *Journal of Development Areas.* v.1 1966, 4/year
Western Illinois University

Morgan Hall 232
Macomb, IL 61455
Tel: 309-298-1108
Fax: 309-298-2865
Indexed in: *Abstracts in Anthropology; America: History & Life; Current Contents; Rural Development Abstracts; Social Science Citation Index; Sociological Abstracts.*

667. *Journal of Development Economics.* v.1 1974, 6/year
North Holland Subsidiary of Elsevier Science B.V.
P.O. Box 211. 1000 AE
Amsterdam, The Netherlands
Tel: 31-20-5803911
Fax: 31-20-5803598
Address in USA:
Elsevier Science Inc.
Box 882
Madison Square
New York, N.Y. 10159
Tel: 212-989-5800
Fax: 212-633-3990
Indexed in: *ABI Inform; Current Contents; Rural Development Abstracts; Social Science Index; Social Science Citation Index; World Agricultural Economic & Rural Sociology Abstracts.*

668. *Journal of Development Studies.* v.1 1964, 4/year
Frank Cass & Co. Ltd.
Gainsborough House
11 Gainsborough Road
London, E11 1RS
England
Tel: 081-530-4226
Fax: 081-530-7795
Indexed in: *America: History & Life; Humanities Index; Rural Development Abstracts; Urban Studies Abstracts; Social Science Index; World Agricultural Economic & Rural Sociology Abstracts.*

669. *Journal of International Affaires.* v.1 1947
Columbia University
Journal of International Affaires
Box 4, International Affaires Bldg.
New York, N.Y. 10027
Tel: 212-854-4775
Fax: 212-864-4847
Indexed in: *America: History & Life; Historical Abstracts; Humanities Index; Social Science Index.*

670. *Journal of International Development: Policy, Economics, & International Relations.* v.1 1981, 6/year
Institute for Development Policy and Management
John Wiley & Son Ltd., Journal Division
Baffins Lane, Chichester
Sussex, PO19 1UD
England
Tel: 0243-779777
Fax: 0243-775878
Indexed in: *Rural Development Abstracts.*

671. *Journal of Marriage and the Family.* v.1 1939, 4/year
National Council of Family Relations
3989 Central Avenue, N.E., Ste. 550
Minneapolis, MN 55421-3921
Tel: 612-781-9331
Fax: 612-781-9348
Indexed in: *Current Contents; Humanities Index; Psychological Abstracts; Social Science Index; Sociological Abstracts; Studies on Women Abstracts.*

672. *Journal of Modern African Studies.* v.1 1963, 4/year
Cambridge University Press
Edinburgh Bldg., Shaftesbury Road
Cambridge, CB2 2RU
England
Address in the USA:
Cambridge University Press
40 W. 20th Street

New York, N.Y. 10011
Tel: 212-924-3900
Fax: 212-691-3239
Indexed in: *Abstracts in Anthropology; Current Contents; Historical Abstracts; Rural Development Abstracts; Social Science Index; Social Science Citation Index; World Agricultural Economic & Rural Sociology Abstracts.*

673. *Journal of Southern African Studies.* v.1 1975, 4/year
Carfax Publishing Co.
Abingdon, OX14 2UE
England
Tel: 44-235-555335
Fax: 44-235-553559
Indexed in: *America: History & Life; Historical Abstracts; Rural Development Abstracts; Studies on Women Abstracts; World Agricultural Economic & Rural Sociology Abstracts.*

674. *Latin American Research Review.* v.1 1965, 3/year
Text in English and Spanish
Latin American Studies Association
c/o University of New Mexico
801 Yale N.E.
Albuquerque, NM 87131-1016
Tel: 505-277-5985
Fax: 505-277-5989
Indexed in: *America: History & Life; Anthropological Literature; Historical Abstracts; Social Science Index; Social Science Citation Index.*

675. *Research in Rural Sociology and Development.* v.1 1984
J A I Press Inc.
55 Old Post Road, No 2
Box 1678
Greenwich, CT 06836-1678
Tel: 203-661-7602
Indexed in: *Sociological Abstracts.*

676. *Resources for Feministe Research/Documentation sur la Recherche Feministe* (text in English and French). v.1 1972, 4/year
Ontario Institute for Studies in Education
252 Bloor St. W.
Toronto, ON M5S 1V6
Canada
Tel: 416-923-6641
Fax: 416-926-4725
Indexed in: *America: History & Life; Historical Abstracts; Humanities Index; Sociological Abstracts; Studies on Women Abstracts; Women's Studies Abstracts.*

677. *Review of African Political Economy.* v.1 1973, 4/year
Carfax Publishing Co.
P.O. Box 25
Abingdon, Oxon OX14 3UE
England
Tel: 44-235-521154
Fax: 44-235-553539
Indexed in: *Current Contents; Rural Development Abstracts; Studies on Women Abstracts; World Agricultural Economic & Rural Sociology Abstracts.*

678. *Rural Africana: Current Research in Social Sciences.* Irregular
Michigan State University
African Studies Center
100 International Center
East Lansing, MI 48824-1035
Tel: 517-353-1700
Fax: 517-353-7254
Indexed in: *America: History & Life; Historical Abstract; P.A.I.S.; World Agricultural Economic & Rural Sociology Abstracts.*

679. *Rural Sociology.* v.1 1936, 4/year
Rural Sociology Society
c/o Patrick C. Jobes, Treasurer
Dept. of Sociology, Wilson Hall
Montana State University
Bozeman, MT 59717

Indexed in: *Abstracts in Anthropology; America: History & Life; Historical Abstracts; Rural Development Abstracts; Science Citation Index; Social Science Index; Studies on Women Abstracts; World Agricultural Economic & Rural Sociology Abstracts.*

680. *Signs: Journal of Women in Culture and Society.* v.1 1975, 4/year
University of Chicago Press, Journal Division
5720 S. Woodlawn Avenue, Chicago IL 60637
Tel: 312-753-3347
Fax: 312-753-0811
Indexed in: *America: History & Life; Humanities Index; Historical Abstracts; Social Science Index; Sociological Abstracts; Social Science Citation Index; Women's Studies Abstracts; Studies on Women Abstracts.*

681. *Society and Natural Resources.* 6/year
Taylor & Francis Ltd.
Rankine Road, Basingstoke
Hants RG24 8PR
England
Tel: 0256-840366
Fax: 0256-479438
Indexed in: *Environment Abstracts; World Agricultural Economic & Rural Sociology Abstracts.*

682. *Sociologia Ruralis* (Text in English, French, and German)
v.1 1960, 4/year
European Society for Rural Sociology
Van Gorcum en Co. B.V.
P.O. Box 43
9400 AA Assen, Netherlands
Tel: 31-5920-46846
Fax: 31-5920-72064
Indexed in: *Abstracts of Rural Development; P.A.I.S.; Rural Development Abstracts; Sociological Abstracts; Social Science Citation Index; Studies on Women Abstracts; World Agricultural Economic & Rural Sociology Abstracts.*

683. *Studies in Family Planning.* v.1 1963, 6/year
Population Council
1 Dag Hammarskjold Plaza
New York, NY 10017
Tel: 212-339-0500
Fax: 212-755-6052
Indexed in: *Current Contents; Environmental Index; P.A.I.S.; Social Science Citation Index; Studies on Women Abstracts; World Agricultural Economic & Rural Sociology Abstracts.*

684. *World Development.* v.1 1973, 12/year
Elsvier Science Ltd.
Pergamon, P.O. Box 800
Kidlington, Oxford OX5 1DX
England
Tel: 44-865-843000
Fax: 44-865-843010
Address in the USA:
Elsevier Science
660 White Plains Road
Tarrytown, NY 10591-5153
Tel: 914-524-9200
Fax: 914-333-2444
Indexed in: *Abstracts of Rural Development; P.A.I.S.; Rural Development Abstracts; Social Science Index; World Agricultural Economic & Rural Sociology Abstracts.*

LIST OF JOURNAL ISSUES RELATED TO WOMEN IN AGRICULTURE

685. *Africa Report* 26:2 (1981).

   Devoted almost exclusively to the topic of African women and their substantial, though not fully realized, role in the continent's economic development. Overview of the current status and future needs of African women as well as articles on what the decade of the 1980s will mean for African-American relations.

686. *Agricultural Education Magazine* 47:12 (June 1975).

   Guest Editor: Elissa Walters.

   Theme issue -- Women in Agricultural Education. Describes the opportunities for women as educators and students in the many fields of agriculture.

687. *Canadian Journal of African Studies; La Revue Canadian des Etudes Africaines* 6:2 (1972).

   Guest Editor: Audrey Wipper.

   Brings together articles about women in East, Central, and West African countries. Examines political roles and tactics, marital and family patterns, economic activities, educational aspirations, and what is needed to change women's role in Africa.

688. *Resources for Feminist Research* 11:1 (1982).

   Guest Editors: E.A.(Nora) Cebotarev and Frances M Shaver.

   This issue includes articles on "Women in North American Agriculture" and "Rural Women in Developing Societies." Ten books on the same subjects are reviewed. Information on "Development Work: For Women by Women" is also included.

689. *Sociologia Ruralis* 28:4 (1988).

Guest editor: Sarah Whatmore.

Devoted to the issue of gender relation on the farm, this issue covers articles on "Agricultural Modernization and the Gender Division of Labor," "Decision Making on the Farm," and "Women's Role in Gender Relations." Reviews several books on gender and power.

690. *Women and Politics* 2:4 (Winter 1982).

Guest Editors: Kathleen A. Staudt and Jane S. Jaquette.

Special theme issue on "Women in Developing Countries: A Policy Focus." Guest editors are Kathleen A. Staudt and Jane S. Jaquette. Includes a chapter entitled "Women, Development and Public Policy: A Selected Bibliography," complied by Sandra Danforth.

ELECTRONIC RESOURCES AND INDEXES

691. *Abstracts in Anthropology.* v.1 1970, 8/year.

   Covers a broad spectrum of current anthropological topics from several hundred journals. Man's speech, physiology, artifacts, history, environment, and social relations are described, analyzed, and interpreted. Published by Baywood Publishing Company.

692. *AGRICOLA* - Coverage: 1970 to the present.

   This is a massive database available from the National Agricultural Library in Washington, D.C. We searched it through Dialog, an information access service company. The coverage is worldwide on all kinds of topics related to agriculture. It is also available in paper entitled *Bibliography in Agriculture*.

693. *America: History & Life* - Coverage: 1964 to the present.

   Includes abstracting and indexing of a full range of U.S. and Canadian history, area studies, and current affairs literature. The database corresponds to the printed index of the same name published by ABC-CLIO. This file is also available through Dialog.

694. *CAB Abstracts* - Coverage: 1972 to the present.

   Contains all records in the 26 main abstract journals published by the Commonwealth Agricultural Bureau in England. We searched it through the information access company Dialog. Includes World Agricultural Economic & Rural Development Abstracts, from which most of the information on our topics came. This index is also available in paper format.

695. *Education Index.* v.1 1929.

A cumulative index to educational publications in the English language, includes also monographs and yearbooks, published by the H.W. Wilson Company. Covers information from preschool to higher and adult education as well as teaching methods and curriculum. Selection of periodicals for indexing is accomplished by subscriber vote.

696. *ERIC* - Coverage: 1966 to the present.

The complete database on educational materials from the Educational Resources Information Center. It corresponds to two print indexes: *Resources in Education,* which identifies the most significant education research reports, and *Current Index to Journals in Education,* an index of more than 700 periodicals of interest to the educational profession. We accessed ERIC through Dialog.

697. *Historical Abstracts* - Coverage: 1973 to the present.

Indexes the world's periodical literature in history and the related social sciences and humanities. The database corresponds to the two companion publications: *Historical Abstracts: Part A, Modern History Abstracts (1450-1914)* and *Historical Abstracts: Part B, Twentieth Century Abstracts (1914 to the Present)* produced by ABC-CLIO. Available through Dialog.

698. *Social SciSearch* - Coverage 1972 to the present.

A multidisciplinary database indexing items from over 1,500 social science journals published throughout the world. It covers every area of the social and behavioral sciences, and it corresponds to the printed Social Science Citation Index published by Institute of Scientific Information. Available from Dialog.

Electronic Resources and Indexes 275

699. *Social Science Index.* v.1 1974

A cumulative index published by the H.W. Wilson Company, covers periodicals in the fields of anthropology, area studies, community health and medical care, economics, family studies, international relations, policy sciences, social work, public welfare, sociology and urban studies. There is a separate listing of citations to book reviews.

700. *Sociological Abstracts* - Coverage 1963 to the present.

Covers the world literature in sociology and related disciplines in the social and behavioral sciences. Over 1,600 journals are scanned each year. Corresponds to the printed index with the same name. Available through Dialog.

701. *Studies on Women Abstracts* - Coverage 1983 to the present.

An international abstracting service with major focus on education, employment, women in the family and community, medicine and health, female sex and gender role socialization, literary criticism as well as historical studies. All major international journals and books on the topic of women are scanned. Published by Carfax Publishing Company in the United Kingdom.

702. *Women Studies Abstracts* - Coverage 1972 to the present.

Indexes 30-35 journals specifically related to women's issues. Covers abortion, education, psychology, employment, violence against women, history, literature, art and music, and many other topics. Published by Transaction Periodicals Consortium, for Rush Publishing Co., Inc.

AUTHOR INDEX

(Numbers in index refer to entry numbers)

Abbott, Susan 377, 512
Abell, Helen C. 376
Achouth, Lalith 483
Adams, Jane 1
Adams, Elizabeth Rose 513
Adekanye, Tomilayo O. 397
African Training and Research Center
    for Women Staff, 438
Afshar, Haleh 297
Agarwal, Bina 203, 398
Aimed, Iftikhar 265, 266
Aimed, Mayan Recognition 204
Alberti, Amalia Margherita 514
Allen, Ruth Alice 2
Als, G. 579
Alston, Margaret Mary 516
Amadi, C.T. 123
Angeles-Bernardo, Estelita 517
Ankarloo, Bengt 3
Anker, Richard 264
Anstey, Barbara Eleanor 518
Aranda, Josefina 162
Ardayfio-Schandorf, Elizabeth 618
Arizpe, Lourdes 162
Ashby, Jacqueline A. 163, 619
Asian and Pasific Development Center
    (APDC) 399
Atal, Yogesh 205
Atkeson, Mary Meek 4

Baas, Ettie 620
Bagchee, Aruna 439
Baker, Gladys L. 5
Baksh, Michael 440
Ballweg, J.A. 441
Barberis, Corrado 581

Barbic, Ana 241, 242, 378
Barlett, Peggy F. 164, 400
Barnes, Carolyn 298
Barrett, Hazel 72
Barthez, Alice 580
Bartley, Paula 6
Barton-Cayton, Amy Elizabeth 519
Baser, Heather Jane 520
Bass, Herman M. 442
Bauder, Ward W. 381
Bauman, Hermann 299
Bay, Edna G. 73, 104
Becouarn, M.C. 582
Beers, Howard W. 7
Bell, J.H. 444
Bell, Lloyd C. 443
Bembridge, T. J. 74
Beneria, Lourdes 267, 268, 300, 301
Beoku-Betts, Josephine 401
Bergen, A. 583
Berkowitz, Alan David 521
Berlan Darque, Martine 379
Berlan, Martine 243
Besteman, Catherine 244
Bindocci, Cynthia Gay 621
Binni-Clark, Georgina 8
Bisilliat, Jeanne 584
Blood, Robert O. Jr. 302
Blumberg, Rae Lesser 269
Bokemeier, Janet 380
Bokemeier, Janet L. 165
Borish, Linda J. 9
Borish, Linda Jane 522
Boserup, Ester 75
Bossen, Laurel 303
Boulding, Elise 166

277

Bourque, Susan C. 167
Bowles, Paul 218
Brady, Marilyn Dell 10
Bramsen, Michele Bo 366
Brandth, Berit 304
Bridger, Susan 245
Bronstein, Audrey 168
Brown, Judith K. 305
Brown, Minnie Miller 11
Brown, Katrina 523
Brown, Judith K. 169
Browne, Angela 72
Bruner, Diana 170
Bryson, Judith 76
Buckley, Joan 99
Buffalohead, Pricilla K. 12
Bukh, Jette 77
Bullwinkle, Davis A. 622
Burchinal, Lee G. 381
Burfisher, Mary E. 306
Burge, Penny L. 469
Burra, Nerra 230
Burton, Michael L. 308, 309
Buttel, Frederick H. 307
Buvinic, Mayra 623

Cadieux, Ron 650
Callaway, Barbara 78
Cano, Jamie 445
Canoves Valiente, Gemma 524
Canoves, Gemma 249
Carbert, Louise I. 525
Carew, Joy Gleason 446
Carney, Judith 13
Carney, Judith A. 79
Carr, Marilyn 80, 228
Cashman, Kristin 526
Cebotarev, E. 447
Cebotarev, E.A. 624
Center for Policy and Development Studies 625
Centro de Investigacion y Estudios de la Reforma Agraria (Nicaragua) 171
Cernea, Michael 246
Chakrapani, C. 448
Chaney, Elsa M. 449
Chapman Smock, Audrey 81
Charlton, Sue Ellen M. 270
Chaudhari, N.V. 236
Cheater, Angela 82
Chen, Marty 403
Cheng, Lucie 626
Clark, Barbara A. 83
Cloud, Kathleen O'Donnel 527
Cohen, Yolande 585
Cohen, Marjorie Griffin 14
Colman, Gould 382
Commins, Stephen K. 417
Compton, Lin J. 456
Connelly, Patricia M. 310
Consultative Workgroup on Women in Rice Farming Systems in the Philippines 235
Conte, Christine 172
Conti, Anna 84
Cooper, Barbara E. 450, 451
Copp, James H. 184
Cormier, D. 586
Corrèze, A. 587
Coughenour, Milton C. 311
Craig, Lee A. 15
Craig, R.A. 627
Cram, Barbara Jean 528
Creevey, Lucy E. 85, 86, 452
Croll, Elisabeth J. 206, 271
Cros, M.O. 588
Culaud, H.P. 589
Curry, Charles 453

Danforth, Sandra 628
Davison, Jean 87, 404, 454

Author Index

de Wild, John Charles 159
Deere, Carmen Diana 173, 174, 175, 272, 312, 313, 314, 405, 406
Dejene, Alemneh 455
Dentzer, M.T. 590
Desjardins, Ghislaine 613
Dey, Jennie 88, 315, 407
Dhanakumar, V.C. 456
Dillingham, John M. 457
Dixon, Ruth 316
Dixon, Ruth B. 273
Dixon-Mueller, Ruth 317
Doebbert, Jan 458
Dollahite, David Curtis 529
Donner-Shabafrouz, I. 591
Drum, Sue 459
Due, Jean M. 89, 90, 91, 408, 409

Eghan, Felicia Rosaline 530
Elbert, Sarah 382
Eldredge, Elizabeth A. 92
Ember, Carol R. 274
Endeley, Joyce Bayande Mbongo 531
Ensminger, Jean 410
Ericksen, Julia 16
Evans, Barbara 17
Evenson, Robert 207
Everaet, H. 592, 593

Fairbanks, Carol 18, 19, 20, 629
Faragher, John Mack 21
Faris, Mohamed A. 630
Farnsworth, Beatrice 22
Fassinger, Polly A. 176
Fassinger, Polly Ann 532
Fast, Vera 640
Fedorova, M. 247
Feldman, Rayah 93
Feldstein, Hillary Sims 318, 319
Fera, Darla 631
Ferchiou, Sophie 594

Filippi, Geneviève 608
Filippi, G. 595
Fink, Deborah 23, 24
Finlay, Barbara 177
First-Dilic, Ruza 248
Flora, Cornelia B. 275
Flora, Cornelia Butler 25, 26
Folber, Nancy 276
Fortmann, Louise 94, 632
Fredricks, Anne 27
Fritz, Susan M. 443
Frombach, Hannelore 533

Gamon, Julia A. 461
Garcia-Ramon, Maria Dolores 249
Garett, Patricia 320
Garkovich, Lorraine 380
Garrett, Patricia M. 178
Gasson, Ruth 250, 251, 321, 322
Gebby, Margaret Dow 28
Geisler, Charles C. 411
Geschiere, P. 596
Ghorayshi, Parvin 179
Gielgud, Judy 534
Gilles, Jere Lee 462
Gillespie, Gilbert W. Jr. 307
Gittinger, J. Price 95
Gladwin, Christine H. 96, 97, 98
Gloss, Molly 29
Godwin, Deborah D. 277, 278
Goldschmidt, Walter 288
Gomez, Stella 163, 619
Goody, Jack 99
Gorman, Pat 463
Graham, Donna L. 464
Granie, A.M. 597
Gregg, Ted 465
Grier, Beverly 100
Gronau, Ruben 279
Guerrero, Sylvia 396
Guillou, Anne 598

Guyer, Jane I. 101, 102, 103, 323

Haakenson, Bergine 20
Hafkin, Nancy J. 104, 633
Haggard, Shirley 478
Hagood, Margaret Jarman 30
Haney, Wawa G. 31, 412
Hanger, Jane 105
Hargreaves, Mary M.W. 32, 33
Harris, Evelyn 34
Harry, Indra S. 180
Hart, Gillian 208
Haugen, Marit S. 252
Haugerud, Angelique 106
Hay, Margaret Jean 107
Heil, Mark Takeo 535
Helsloot, L. 599
Hempstead, Katherine Ann 536
Henderson, Janet L. 450, 451
Henn, Jeanne Koopman 108
Heppner, Paul 280
Herz, Barbara 109
Herzog, W. 600
Hetland, Per 253
Hewitt, Mary 466
Heybroek, Hans M. 36
Heyer, Judith 110
Heyzer, Noeleen 209
Higgins, Kathleen Mansfield 467
Hill, Kim 113
Hill, Frances 468
Hill, Polly 111
Hill, Bridget 35
Hillison, John H. 469
Hirchmann, David 112
Holmes, Francis W. 36
Hombergh, Heleen van den 324, 634
Horenstein, Nadine R. 306
Howell, David L. 466
Huffman, Wallace E. 325, 326
Huntington, Suellen 281

Hurtado, Magdalena A. 113
Hussain, Sahadad M.D. 210

Ilyas, S.A. 470
International Labour Organization 114, 282
International Bank for Reconstruction and Development 327
Ireson, Carol 211

Jacobson, Doranne 212
Jacoby, Hanan G. 328
Janiewski, Dolores 37
Jaquette, Jane S. 431
Jarosz, Lucy Antonia 115, 116
Jeay, Anne-Marie 617
Jeffrey, Julie Roy 38, 39
Jellison, Katherine Kay 537
Jensen, Joan M. 40, 41, 42, 43, 44, 45, 46
Jiggins, Janice 283, 635
Johnson, Patricia Lyons 214
Johnson, Marshall 213
Joint Bank-Fund Library 636
Jones, Lu Ann 47
Jones, Calvin 181
Jordan, Stephen A. 538
Joyce, Lynda 637
Judd, Ellen 215
Junzuo, Zhang 491

Kabeer, Naila 413
Kaberry, Phyllis M. 117
Kada, Ryohei 284
Kala, C.V. 329
Kalb, Marion 131
Kandiyot, Deniz 414
Kardam, Nuket 415
Keating, Nora 330
Keim, Ann Mackey 359, 385
Keller, Bonnie 416

Author Index

Kessler, Shelly 638
Khalil, Ahmed A. 470
Khan, Mahmood Hasan 630
Khan, Zarina Rahman 216
King, Evelyn 639
Kinnear, Mary 48, 640
Kinsey, J. 471
Klein, Gary 16
Kleinegger, Christine Catherine 539
Klepper, Betty 472
Knotts, Don 474
Knotts, Rose 474
Knowles, Jane B. 31
Knudsen, Barbara 182
Koenig, Dolores 118
Kohl, Seena B. 331
Kolde, Rosemary F. 473
Kossoudji, Sherrie 119
Kuehl, R.J. 475
Kumar, K. 476
Kurian, Rachael 49
Kuznik, Anthony 477
Kwafo-Akoto, Kate 618

Lagrave, Rose-Marie 601, 602
Lakshmi, Bharadwaj K. 393, 395
Lancaster, C.S. 332
Lapido, Patricia 333
Lasram, M. 603
Lastarria-Cornhiel, Susana 334
Lawrence, Roger Lee 540
Leacock, Eleanor 335
Leadley, Samuel M. 637
Leckie, Gloria J. 336
Leckie, Gloria Jean 541
Lee, Jasper S. 47
Lee, Delene W. 478
Lefaucheux, Marie-Helen 120
Leibelt, Don C. 480
Lele, Uma 285
Leon, Magdalena 406

Leon de Leal, Magdalena 175, 314
Leplaideur, S. 604
Leske, Gary 481
LeVine, Robert A. 121
Lewis, Barbara C. 286
Lewis, Martha Wells 641
Li, L. 441
Lindemann-Meyer zu Rahden, H. 605
Lindsay, Beverly 482
Lofchie, Michael F. 417
Long, Norman 338
Lopez-Trevino, Maria Elena 542
Lothrop, Gloria Ricci 41
Loutfi, Martha Fetherolf 337
Loxton, Cathy 6
Lucas, Kimberley 543
Lupri, Eugene 392
Lynch, Barbara Deutsch 183
Lyson, Thomas A. 217

MacDonald, Martha 310
Machum, Susan T. 544
Mackenzie, Fiona 341
MacPhail, Fiona 218
Magayane, F. 408
Mago, Sneh Lata 476
Mandala, Elias 339
Manjula, N. 483
Manyeh, Marie Aliena 545
March, Candida 368
Maresca, Sylvain 606
Maret, Elizabeth 50, 184
Margai, Magdalena 546
Marlowe, Julia 277
Martelet, Penny 51
Marti, Donald B. 52
Martin, Susan 122
Martin, S. 607
Mascarenhas, Ophelia 642
Massiah, Joycelin 340
Matsumoto, Valerie Jean 547

Mayoux, Linda 342
Mbata, J.N. 123
Mbewe, Dorcas Chilila 416
Mbilinyi, Marjorie 124, 642
Mbilinyi, Marjorie J. 125, 484, 485
McCandless, L.Z. 506
McCurry, Dan C. 53
McDonald, Mark B. 192
McHenry, Dean E. Jr. 126
McMillan, Della 96
McMurray, Sally 54
McSweeney, Brenda Gael 287
Meera, B. 486
Mehenna, Sohair 263
Meiners, Jane Ellen 548
Mencher, Joan P. 219, 220
Michaelson, Evelyn Jacobson 288
Mickelwait, Donald R. 289
Mies, Maria 290
Mikell, Gwendolyn 127, 128
Miller, Lorna Clancy 412
Miller, Barbara D. 221
Mincolla, Joseph Anthony 549
Minge-Kalman, Wanda 254
Mitra, Manoshi 55
Mkandawire, Richard M. 129, 130
Moerkeberg, Henrik 255
Monson, Jamie 131
Montgomerie, Deborah 343
Moock, Joyce Lewinger 344
Moock, Peter R. 418
Moore, Keith M. 345
Moore, Henrietta 56
Morandy Nostrakis, Anny 550
Morris, Jon 105
Morrison, Denton E. 391
Morvaridi, Behrooz 419
Moser, Caroline O.N. 420
Muhl, H. 600
Munro, Brenda 330
Murray, Collin 132

Mwaniki, Nyaga 346

Nair, G.T. 486
Nelson, Nici 133, 222, 643
Newbury, Catherine M. 421
Nicourt, C. 595
Nicourt, Christian 608, 609, 610
Norris, Mary E. 291
Nwanosike, Eugene O. 644

Obidi, S.S. 487
Odie-Ali, Stella 185
Okelo, Mary 134
Okeyo, Achola Pala 488
Olayiwole, Comfort Bang 551
Olenja, Joyce M. 135
Ollenburger, Jane C. 186
Olmstead, Judith 136
ORSTOM, Laboratoire de sociologie et géographie africaines 611
Orvis, Stephen 137, 422
Osterrud, Nancy Grey 47, 57

Pala, Achola O. 138
Palmer, Ingrid 292
Pandey, U.S. 444
Pankhurst, Donna 347
Panter-Brick, Catherine 223
Paret, Andrea Martha 489
Parson, Karen 552
Pausewang, Siegfried 423, 424
Pearson, Jessica 348
Pescatello, Ann 187, 645
Peters, Pauline 349
Petronoti, Marina 256
Pfeffer, Max J. 350
Phillips, Anne Radford 553
Pickett, Evelyn L. 554
Plaza, P. 603
Poats, Susan V. 318, 351
Price, Lisa Marie Leimar 555

Author Index

Quade, Ane Marie 556
Quisumbing, A.A. 490

Racine, Philip N. 58
Radcliffe, Sarah A. 352
Rafats, Jerry 646
Rai, Shirin M. 491
Rakodi, Carole 139
Ram, Rati 492
Randolph, Sheron 425
Rattin, S. 612
Rea, Jennette 493
Rees, Josephine Duggan 59
Reimer, Bill 353
Reitz, Karl 308
Reynolds, Carl L. 494
Reynolds, Lucile W. 557
Rialland-Morisette, Yvonne 613
Riegelhaupt, Joyce F. 257
Rihani, May 647
Roberts, P. 614
Roberts, Penelope A. 140
Roberts, Sarah Ellen 60
Rogers, Susan 383
Rogers, Barbara 426
Rome, Roman Albert 558
Rose, Margaret 61
Rose, Margaret Eleanor 559
Rosenfeld, Rachel A. 181, 188, 364, 384
Ross, Lois L. 189
Rossier, R. 615
Rossini, Rosa Ester 190
Rothstein, Frances 354
Roux, P. 597
Ruchwarger, Gary 191
Russel, Scott C. 192

Sabol, Lois Malmberg 560
Sachs, Carolyn E. 193, 194, 355
Sackville-West, V. 62

Safa, Helen I. 335
Safilios-Rothschild, Constantina 427
Sagrario Floro, Maria 356
Saito, Katerine A. 141, 142, 293, 357
Sajogyo, Pudjiwati 224
Salamon, Sonya 358, 385
Sanday, Peggy R. 294
Sanders, Rickie 425
Saradamoni, K. 220, 225
Sato, Kunio 226
Sawer, Barbara J. 386
Schipper-Peters, Caroline 561
Schmidt, Elizabeth 63
Schoustra-van Beukering, E.J.E. 227
Schumacher, Ilsa 295
Schwartz, A. 616
Schwarzweller, Harry K. 176
Schwieder, Dorothy 64, 496
Scott, Gloria L. 228
Secretariat of the COPA Women's Committee 258
Seever, Brenda 497
Seger, Karen Elizabeth 562
Sen, Gita 267
Shapiro, David 143
Sharma, D.K. 387
Sharma, Ursula M. 229
Sharpless, Mary Rebecca 563
Shaver, Frances M. 66, 195
Sheehan, Nancy 359
Shortall, Sally 196, 388
Shorter, Frederic C. 259
Sibalwa, David 498
Siddaramaiah, B.S. 483
Simons, Thordis 65
Simpson, Ida Harper 360
Singh, Ram D. 492
Singh, Tej Ratan 387
Singh, Andrea M. 230
Sirisena, N.L. 232
Smith, Charles D. 144

Smith, Deborah Beth 564
Smith, Rosslyn Braden 565
Smith, Suzanna 499
Smith-Sreen, John 231
Smith-Sreen, Poonam 231
Smuksta, Michael J. 566
Sontheimer, Sally 296
Souron, Olivier 610
Sousi-Roubi, Blanche 260
Spring, Anita 145, 146, 147, 148, 361
Sproles, Elizabeth Kendall 500
Spurling, Daphne 293
Standing, Guy 501
Stanley, Autumn 67
Starr, Karen 68
Statigaki, Maria 362
Staudt, Kathleen 428, 429, 430, 431, 432, 433, 434
Staudt, Kathleen A. 149, 150, 151
Staudt, Kathleen Ann 567
Steele, Roger E. 502
Stephens, Alexandra 503
Stevens, Hélène 617
Stevens, Lesley 144
Stichter, Sharon 107
Stitz, John 25, 26
Stoeckel, John 232
Stoler, Ann 233
Stone, Margaret Priscilla 568
Stratigaki, M. 261
Straus, Murray A. 197, 198, 363
Stuart, Robert C. 435
Sturgis, Cynthia 69
Sudha, Rani P. 436
Sukkary-Stolba, Soheir 262
Sundberg, Sara Brooks 18, 629
Susheela, H. 234
Swanson, Louis 311
Sweet, James A. 199
Swindell, Ken 152
Szychter, Gwendolyn Mary 569

Taylor, Barbara 499
Thomas, John K. 504
Thompson, Barbara 505
Thompson, O.E. 506
Thorpe, Trevor A. 180
Tigges, Leann M. 364
Tima, Grace Ngemukong 570
Timber, Priscilla Jean Tomich 571
Tinker, Irene 365, 366, 648
Toth, James 367
Tripp, Robert B. 153
Tshatsinde, M.A. 154
Tupas, Lourdes Pedregosa 572

University of West Indies, Barbados 649

Valentine, Frances Wadsworth 200
van Es, J.C. 394
Van Allen, Judith 155
Vangile, Titi 573
Vaughan, Megan 56, 112
Vavrus, Linda Gire 650
Vellenga, Dorothy Dee 156
Vetter, Louise 508
Viola, Lynne 22
Vlassoff, Carol 509

Waghmare, S.K. 236
Wagner, Maryjo 574
Walker, Tjip S. 157
Wallace, Tina 368
Wangari, Esther 575
Warren, Barbara Kay 167
Watts, Michael 13
Webb, Anne B. 70
Weil, Peter M. 158
Weintraub, Irwin 651
Whatmore, Sarah 369, 370
White, Chrisine Pelzer 237
White, Douglas R. 309, 371

White, Marcia 90
Whitehead, Vivian B. 652
Whiteley, Ellen H. 459
Whittaker, Wesley Lloyd 576
Whittington, Susie 510
Wilkening, Eugene A. 389, 390, 391, 392, 393, 394, 395, 396
Wilson, John 372
Wilson, Fiona 201
Wilson-Larson, Laurie 577
Wilson-Moore, Margot 238
Wipper, Audrey 160, 161
Wood, Lucie Saunders 263

Xiyi, Huang 239

Yates, Barbara A. 182
Young, Kate 202, 373, 374, 375
Youssef, Nadia H. 240

Zambia Association for Research and Development (ZARD) 511
Zappi, Elda Gentili 71, 578

SUBJECT INDEX

(Numbers in index refer to entry numbers)

Abstracts in Anthropology 691
access to education 484
access to land 438
afforestation project 573
Africa 351, 614, 633
Africa Report 685
African Development Bank 134
African Studies Center at Michigan State University 161
African women 96
agrarian feminism 525
agrarian reform 320, 405
agrarian reform law 171
agrarian reform program 334
agrarian structure 282
agribusiness 124, 162, 480, 537
agribusiness education 505
AGRICOLA 692
Agricultural Adjustment Act 11
agricultural change 102
Agricultural Change, Rural Women and Organization 399
agricultural development projects 306
Agricultural Economics 653
Agricultural Education Magazine 654, 686
agricultural employment 316
agricultural extension 293, 437, 439, 462
agricultural extension for women 157
agricultural extension work 422
Agricultural History 655
agricultural intensification 309
agricultural labor 263
agricultural labor force 286
agricultural machinery 442
agricultural policy 341, 406

Agricultural Producer Cooperatives 246
agricultural production 193, 194, 229, 272, 285, 293, 353, 416, 491
agricultural productivity 428
agricultural research and extension 319
agricultural support services 182
agricultural teachers' attitude 445
agricultural techniques 75
agricultural technology 270, 283, 446, 635
AICRP Home and Farm Management 234
Akan lineage principles 127
Alabama 65
Alberta 32, 60
Amenagement des Valles des Volta (AVV) 84
America History & Life 693
American Anthropologist 656
American Journal of Agricultural Economics 657
Amish 16
Andes 167, 301, 314
Angola 80
animal husbandry 145, 192, 236, 239, 263, 309, 442
Anthropological Quarterly 658
Arizona 192
Asia 351
Association of African Women for Research and Development 114
audiovisual training 590
Australia 444, 516, 627
Austria 335

autocorrelation analysis 371

Bamako 118
Bamako Workshop 86
Bangladesh 204, 205, 208, 210, 216, 222, 228, 238
BaSotho society 92
bee-keeping 631
behavioral model 325
Belgium 258, 2960
Bemba 169
Bengali women 227
Beynon, Francis Marian 17
Bihar 55
biological reproduction 297, 300
black agriculturists 651
black women 11
Blood and Wolfe's testing technique 377
Bolivia 168, 289
Boone County 23
Boserup, Ester 267, 281, 308, 309, 312, 365
Botswana 80, 94, 119, 318, 349, 467
Brazil 190
Brooks, Sara 65
Buisman, Christine Johanna 36
Burkina Faso 142, 318, 492, 604
butter making 40, 42, 57, 552, 631

CAB Abstracts 694
California 41, 442, 519
class and sex 430
Cameroon 101, 102, 108, 117, 157, 323, 499, 531, 570, 596
Canada 8, 14, 18, 60, 66, 170, 179, 195, 196, 330, 336, 376, 386, 525, 544, 546, 569
Canadian Council on Rural Development (C.C.R.D.) 170
Canadian Farm Women's Network 196
Canadian Journal of African Studies/ La Revue Canadian des Etudes Africaines 659, 687
canning industry 53
career development 504
career goals 472
career opportunities 475
career options workshop 280
Caribbean 182, 240, 351, 649
cash crop production 273
cash-crop economy 339
cash economy 297
cassava 421
Catalonia 249
cattle ranching 331
cattle trail 639
census of agriculture 326
Central People's Government 239
Chavez, Helen 61
cheese 54, 57
Chewa peasantry 129
children 379
Chile 178
China 206, 209, 213, 215, 239, 271, 301, 399, 491, 507, 626
civil rights 45
class dynamics 347
class relations 369
cocoa 100, 127, 128, 156, 323,
coffee 202
Colden, Jane 493
Colombia 175, 318, 340
Colorado 166, 348, 574
Columbia Basin Project 197
communal farming 126
community adjustment 205
Comstock, Anna Botsford 493
Concerns-Based Adoption Model (CBAM) 526
Connecticut 9, 522

Subject Index

contract farming 115
cooking energy crisis 296
cooperatives 159, 237, 248, 334, 342, 362, 378
COPA Women's Committee 258
coping 529
coping styles 280
Cornell University 67
Cornell's University's Farm Family Project 382
Cortez Colony 547
cotton farms 563
cotton 2, 202, 339
Country Life reformers 536
credit 185, 295
credit systems 357
crop pattern 329
crop production 236
crop tending 308
crop type 371
cross-cultural models 281
Cuba 271
Current Anthropology 660

dairy farmers 245
dairy farming 54, 231, 327
dairy products 556
Dakota 32
data collection methods 264
decline 312, 355
Denmark 255, 258, 260
Detroit 302
Development Alternatives Inc. 289
Development and Change 661
development management 434
development planners 361
development plans 160
development programs 366, 368, 614
development strategies 215, 270, 374
diaries 29

Division of Labour Module 81, 298
divorce 77
domestic labor 310, 342, 539
domestic space 379
domestication of women 426
domesticity 1
Dominican Republic 177
Dutch Elm disease 36

East Africa 484, 485, 513
Economic Development and Cultural Change 662
economic diversification 367
Ecuador 168, 514
education 267
Education Index 695
Egypt 262, 263, 367, 630
El Salvador 168, 334
emancipation 242
employment migration 441
employment patterns 199
empowerment 416, 417
energy problem 118
England 35, 54, 369, 556
equal opportunities 429
Equal Remuneration Act 436
equal wages 429
equality 120
ERIC 696
Ethiopia 136, 301, 424, 455
ethnic groups 41, 26
ethnicity 514
ethnographic field research 432
Ethnology: An International Journal of Culture and Social Anthropology 663
Evans, Alice Catherine 493
extension agents 497
extension education 464
extension home economists 551

extension programs 572
extension services 140, 148, 180, 293, 570, 344, 357, 588
external markets 213

family business 250, 251
family enterprise 369
family farming 345
family farming system 370
family farms 184, 338
family food production 641
family income 354
family planning 366
family production system 240
family relations 278
family relationships 389
Farm and Home 4
farm business 387
farm entrepreneurs 478
farm family 7
Farm Family Test Group 539
farm homes 376
farm households 307
farm improvement 396
farm labor 353
farm labor requirements 348
farm operation 176
farm operators 198, 389, 390
farm ownership 593
farm production 311
farm size 360
farm tenure 558
farm wives 226, 243, 255, 277, 345, 369, 386, 389, 608
farm women 21, 165, 166, 189, 241, 468
farm women - health 9
farm women - writings 20
Farm Women's Survey 188, 384
Farmers' Alliance 574
Farmer's Wife 592

farming households 244
farming practice 26
farming system research 344
farming systems 89
farming systems and extension 147
farming technology 526
farmland ownership 411
federal tax law 45
Federal Republic of Germany 350
female artisans 431
female enrollment 500
female farm operators 541
female farmers 602
female-headed households 119, 265, 408
female invisibility 269
female land ownership 385
female/male interaction 383
female operated farms 244
female roles 288
female seclusion 221
female subordination 300
female wages 291
feminist awareness 431
feminist perspective 516
feminist theory 370
feminization of agriculture 246
feminization of production 350
field-work techniques 264
Finland 338
fisherman farmers 253
fishing community 310
Food Corps Program International (CILCA) 85, 86
food crisis 97, 103, 150, 296
food crops production 123
food production 101, 148, 650
food security 88, 95, 134
food survey 223
forage feeding 615
forest 296

Subject Index

forest work 211
forestry 239, 327
France 258, 260, 589, 595, 597, 601
frontier women 38
fruit growing 53
Future Farmers of America (FFA) 466, 469, 506

Gallon's problem 371
Gambia 72, 79, 152, 158, 315, 407
gender analysis 319, 322
gender-based inequalities 266
gender, environment and development (GED) 324, 634
gender equity 527
gender differences 280, 292, 304
gender hierarchies 300
gender inequality 335
gender issues 318
gender policy 420
gender relations 336, 352
Georgia 164
Germany 258, 260
Ghana 77, 100, 111, 127, 128, 153, 156, 289, 340, 618, 530
global crisis 368
government policy implementation 428
grain 556
grain farming 331
Grain Grower's Guide 17
Grange, the 52, 57
Greece 256, 258, 261, 362
green certificate program 507
green revolution 301
Guatemala 168
Guyana 185

Harris, Emily Lyles 58
Hausa women 78
health 538

health care 327
heavy agricultural machinery 304
Heraklion, Crete 261
hierarchies of gender 140
Himachal Pradesh 229
Hindu women 238
Historical Abstracts 697
hoe culture 299
home economics 64, 517, 540
homestead agriculture production 210
homestead gardening 238
horticultural production 72
household as a unit analysis 373
household development 276
household economy 254
household income 89
household labor 15
household production 42, 268
household tasks 302
household strategies 244
household structure 214
household topologies 349
household work activities 548
Huerta, Dolores 61
Human Organization 664
Hungary 242
hydroelectric dams 209

Illinois 1, 358, 385, 566, 576
impact model 201
independent farmer 364
India 55, 203, 205, 209, 212, 220, 221, 222, 225, 229, 231, 236, 290, 296, 329, 387, 398, 399, 403, 448, 456, 476, 509
Indonesia 218, 318, 399
inequality 488
inheritance 289
insecticide safety 518
Integrated Rural Survey 81
integration of women 320

intensification of agriculture 274
intergenerational transfers 490
International Conference of Ethiopian Studies 424
International Council on Women 120
International Labour Organization (ILO) 282
interview guides 319
intra-household dynamics 351
invisible farmers 286
Iowa 3, 24, 64, 325, 326, 381, 496, 518, 560
Iowa State College 64
Ireland 196, 251, 258, 260, 388
Iroquois 169
irrigation 183, 209
irrigation projects 79, 407, 591
Islamic countries 297
Italian feminism 71
Italy 71, 258, 260, 578, 581

Jamaica 449
Japan 226
Japanese Americans 547
Java 208, 224, 233
Journal of African History 665
Journal of Development Areas 666
Journal of Development Economics 667
Journal of Development Studies 668
Journal of International Affairs 669
Journal of International Development: Policy, Economics, & International Relations 670
Journal of Marriage and Family 671
Journal of Modern African Studies 672
Journal of South African Studies 673

Kansas 25, 574
Kansas Farmers' Alliance 10

Kansas Knights of Labor 10
Kentucky 311, 380
Kenya 93, 105, 106, 109, 110, 115, 135, 137, 142, 149, 298, 318, 341, 346, 377, 401, 404, 418, 422, 430, 454, 523, 543, 567, 575
Kerala 225, 486
Kerling, Louise Catharine Petronella 36
Kikuyu women 402, 454, 512
Kisii 137
Kofyar farming system 568
Korea 205
Krishi Vigyan Kendra (KVK-Farm Science Centers) 456

labor demand 200
labor force 501
labor force participation 165, 186, 190, 264, 291, 436, 473, 564
labor force statistics 268, 273
labor market statistics 277
labor tenure 159
Lady Ann Vavasour 67
land holdings 234, 316
land ownership 185, 225, 275, 358
land rights 398
land tenure 329
Land Tenure Center (LTC) 359
land tenure reform 106
land tenure patterns 385
land tenure policies 404
land tenure system 178
land use 87, 296, 407
land use patterns 75, 359
land use priorities 159
land usage 373
Land Grant Colleges 494
land-grant universities 450, 451
Laos 211

Subject Index

Latin America 163, 187, 201, 296, 351, 405, 406, 447, 619, 645
Latin American Research Review 674
leadership training 481
Ledeboer, Maria Sara Johanna 36
legal rights 317
legal status 289
leisure-time activities 557
Lesotho 80, 92, 132, 289, 349, 427
Lewis, Faye Cashatt 32
Lilongwe rural development project 129, 145
lima recommendations 520
literacy 491
Lithuania 561
livestock judging 481
local conditions 361
local power structure 149
log-linear analysis 308
low productivity 154
Luxembourg 258, 260

machine knitting 431
machine operators 245
Madagascar 116
maize 418
maize production 237
Malawi 80, 83, 112, 129, 130, 145, 146, 147
Malaysia 209, 399
male dominance 288, 358
male-female relationship 485
Mali 85, 86
management instruction 458
MANCOVA 278
Mandinka gender relations 13
Manila 226
Manitoba 48, 640
Manu River Basin 113
Marble, Delia W. 51
market agriculture 568

market orientation 316
market production 265
marketplace 297
Marriage Contract Law 125
marriage payments 132
Maryland 34
mass media 565
matrilineal society 112
McCormick, Fannie 10
Mediterranean 603, 617
Mediterranean women 256
Mexican American women 542
Mexico 162, 202, 301, 354, 373
Mexico City 366
Michigan 532
Michigan Income Dynamics 279
Middle East 351, 367
migrant labor 132
milking 615
Minnesota 12, 70, 458, 564
Mississippi 38
models of development 138
modern farm practice 571
modernization of agriculture 195, 303
Mormon village 69
Mozambique 80
Muslim women 238, 240

Namiele project 616
Nanking (Jingling) University 213
Nanticoke Valley 57
National Agricultural Library 652
national development program 232
National Conference on American Farm Women 31
National Opinion Research Center (NORC) 181, 384
national parks 513
national survey 181
Native American women 45, 46
Native Americans 6

Navajo 192
Navajo reservation 172
Nebraska 23, 443, 574
Nemow River 292
Nepal 223
Netherlands 36, 258, 260
Nevada 554
New England 54
New Guinea 214, 503
New Mexico 43, 45
New York State 7, 54, 57, 307, 360, 382
New Zealand 217, 343
Nicaragua 171, 191, 342, 535
Nicaraguan Agrarian reform 174
Nigeria 78, 122, 123, 142, 289, 301, 306, 323, 333, 397, 487, 526, 551, 568
Nonpartisan League 68
nontraditional employment 510
nontraditional opportunities 508
Norman Girvan's categories 520
North Carolina 30, 37, 325, 326, 372, 553
northern plains 33
Norway 252, 253, 304
Nova Scotia 310

OEF International 131
off-farm earnings 277
off-farm employment 193, 278, 284, 293, 307, 311, 321, 345, 372, 380, 544, 576
Ohio 28, 54, 445
Ojibway Indian 12
Oklahoma 166, 325, 326
Oklahoma State University 489
Olgivi, Ida H. 51
Ontario 14
oral history 47
Orma women 410

Pack, Emma D. 10
Pakistan 222
palm production 122
Paraguay 289
part-time farming 164, 284, 350
Patch, Edith Marian 493
patriarchal structure 362
patriarchy and class 140
patrilineal community 135
patrilineal extended family 227
patrilineal inheritance 99
peanuts 421
peasant household production 124
peasant migration 352
peasant women 383
Pennsylvania 44, 57, 442
Peru 167, 168, 173, 175, 183, 289, 312, 328, 352
Peruvian Living Standard Survey 328
pesticides 518, 542
Philadelphia 40, 44
Philippines 205, 207, 235, 318, 356, 399, 490, 572, 625
Pickney, Elizabeth 493
pioneer family 60
plantation system 49
plow agriculture 308, 309
pluriactivity 253
policy planners 365
political relationships 434
polygyny 371
Populist Party 574
Portugal 244, 257, 258
Potter, Beatrix 493
poultry 57
poultry keeping 236
power 383
power and negotiation 379
power relations 324, 413
power relationships 388
prairie frontier 18, 629

prairie women 19
primatologists 513
Problem Solving Inventory 280
production cycle 275
production strategies 400
productive capital 410
productivity 294, 295, 492
professional farmers 252
public policy 628

Quebec 66, 179, 195, 586, 613
questionnaire design 264

ranching industry 554
random sampling technique 210
random visit method 153
Research in Rural Sociology and Development 675
research on women farmers 433
research theories 637
Resources for Feminist Research/ Documentation sur la Recherche Feminist 676, 688
Review of African Political Economy 677
Rhodesia 169
rhythm of the biological process 275
rice cultivation 219
rice farming systems 235
rice fields 71, 578
rice development project 292
rice development schemes 315
rice irrigation 583
rice production 158, 208, 218, 233, 237,
rice settlement 105
rice villages 490
right of succession 252
Roberts, Sarah Ellen 32
role analysis 560
role conflict 521

role-playing exercises 434
Romania 246
Rural Africana 161
Rural Africana: Current Research in Social Sciences 678
rural development 282
rural farm wives 199
Rural Income Distribution Survey (RIDS) 119
rural management 435
rural migrant 205
Rural Sociology 679
rural women's work 337
Russian peasant women 22
Rwanda 425

SADCC 80
Sahel 85, 86, 452, 550
Saskatchewan 8, 331
Saskatchewan Farm Movement 17
Saskatchewan Women Grain Grower's Association 17
Schwarz, Marie Beatrice 36
sectoral economic planning 403
self-help groups 346
self-identity 380
self-management 599
Senegal 114, 599
Senegal River Valley 583
Senegambia 13
sharecropping 116
shea nuts 604
Shona women 63
Sierra Leone 427, 545
Signs: Journal of Women in Culture and Society 680
silk culture 631
Singapore 301
slavery 371
Slovenia 241
small-scale farming 575

small-scale technology 366
smallholder agriculture 110, 137, 314, 316
smallholder farmers 543
smallholder households 144
social power 233
Social Sciences Index 699
Social SciSearch 698
social security provision 589
socialist development strategies 271
Society and Natural Resources 681
socioeconomic status 373
Sociologia Ruralis 682, 689
Sociological Abstracts 700
South Africa 130, 154
South Asia 643
South Carolina 34, 58
Southern Farm Alliance 39
Soviet Union 22, 245, 271, 435
Spain 258, 524
specific factor analysis 291
Spierenburg, Barendina Gerarda 36
Sri Lanka 49, 222, 232, 399
St. Lucia 182
Stanford University 67
Stanley, Dr. Louise 5
state farm 191
statistical data collection 374
status of women 294, 471
Stevens, Ernestine 5
strange farmer system 152
strawberry export 162
Stress 533
Strong, Harriet Williams Russel 493
structural adjustment plans 356
Structural Adjustment Program (SAP) 98
structural transformation 285
Studies in Family Planning 683
Studies on Women Abstracts 701
sub-Saharan Africa 76, 121, 133, 141, 142, 155, 332, 409, 583
subordination 375
subsistence activities 305
subsistence agriculture 313
subsistence farming 419
subsistence production 268, 375
survey techniques 287
sustainable development 340
Swaziland 80
Sweden 3
Switzerland 254, 615

Tanzania 80, 89, 91, 108, 144, 124, 125, 126, 271, 431, 642
task involvement 395
task participation 577
task performance 381
tea cultivation 237
technological change 265, 266, 363, 400, 419
technology 228, 293, 295
technology transfer 344
telephone survey 307
tenant farm women 30
tenure regimes 359
Texas 2, 457, 563
Texas beef cattle industry 50
textbook 107, 434, 463
Thailand 399, 555
The Farmer's Wife 4, 10, 539
Third World 632
Third World agriculture 317
Third World women 269, 638, 644
Timber Corps 59
time allocation 207, 287
time budget studies 279
time use 440, 548
time allocation 609
tobacco 37, 191, 553
Tobit analysis 277
tools 113

Subject Index

tools, research 219
training 317
Transkei 74
tribal women 236
Trinidad 180
Triple Role Framework (TRF) 413
tropical Africa 159, 417
tropical agriculture 143
Tunisia 340, 594
Turkey 259

UCLA African Studies Center 131
Umoja Federation 149
UNESCO 287
United Farm Women 48
United Farm Workers (UFW) 61, 528, 559
United Kingdom 250, 251, 258, 260, 534
United National Agencies 495
United Nations 269
United States 193, 194, 200, 355, 411, 471, 558, 632
United States Agency for International Development (USAID) 146
United States Department of Agriculture 5, 188
University Center of Dschang 499
University of Illinois 73
University of Minnesota Technical College 477
University of Wisconsin 67
Upper-Volta 84
urban agriculture 139
urban labor markets 267
USDA Forest Service 549
Utah 69

Vermont 166
veterinary medicine 459
Vietnam 237

vocational training 247, 600

wage differentials 436
Wallace's Farm Journal 496
Washington (state) 53
waste disposal 615
wasteland development 230
water management 296
Ways of Coping Scale 280
weavers 136, 202
Wellesley College 178
Went, Johanna Catharina 36
West Germany 392, 394
Westerdijk, Johanna 36
Wife Role Supportiveness Index 363
wild plant food 555
Willoughby, Frances 67
Wilson, James 5
Wisconsin 198, 345, 363, 389, 390, 391, 392, 393, 396, 565
Wisconsin Survey Research Laboratory 393
Women and Politics 690
women farmers 156, 217, 258, 293, 567, 610
women homesteaders 70
Women in Agricultural Development (WIAD) 351
Women in Development (WID) 414, 415
women in management 245
Women Involvement in Farm Economics (WIFE) organization 412
women power 257
women settlers 262
Women Studies Abstracts 702
women suffrage 71
women workshop 375
Women's Bureau of the Department of Labor 53

women's careers 585
Women's Committee of the African Studies Association 104
women's cooperatives 133, 333
women's economic status 203
women's farm work 330
women's inventions 67
women's invisibility 427
Women's Land Army 27, 51, 59, 62
Women's Land Service 343
Women's Liberation Movement 170
women's movement 365
women's network 172
women's participation in agriculture 312
women's program 515
women's status 335, 448, 509, 587
women's training 452
women's workload 262
work roles 176
World Bank 141, 142
World Development 684
World Employment Program 290
Wyman, Walker D. 32

Xhosa society 74

Yemen Arab Republic 562
Yugoslavia 248, 378

Zaire 143, 421
Zambia 56, 80, 89, 90, 139, 142, 318, 416, 423, 498, 511, 520,
Zimbabwe 63, 80, 82, 95, 347, 573